The Artificial Ape

The Artificial Ape

How Technology Changed the Course of Human Evolution

Timothy Taylor

THE ARTIFICIAL APE

First published in 2010 by
PALGRAVE MACMILLAN®
in the United States—a division of St. Martin's Press LLC,
175 Fifth Avenue, New York, NY 10010.

Where this book is distributed in the UK, Europe, and the rest of the world, this is by Palgrave Macmillan, a division of Macmillan Publishers Limited, registered in England, company number 785998, of Houndmills, Basingstoke, Hampshire RG21 6XS.

Palgrave Macmillan is the global academic imprint of the above companies and has companies and representatives throughout the world.

Palgrave® and Macmillan® are registered trademarks in the United States, the United Kingdom, Europe, and other countries.

ISBN: 978–0–230–61763–6

Library of Congress Cataloging-in-Publication Data

Taylor, Timothy, 1960 July 10–
The artificial ape : how technology changed the course of human evolution / Timothy Taylor.
p. cm.
Includes bibliographical references and index.
ISBN 978–0–230–61763–6 (hardback)
1. Human evolution. I. Title.

GN281.T39 2010
599.9398—dc22 2010007924

A catalogue record of the book is available from the British Library.

Design by Newgen Imaging Systems (P) Ltd., Chennai, India.

First edition: July 2010

10 9 8 7 6 5 4 3 2

Printed in the United States of America.

For my daughters, Rebecca and Josephine

Creatures of a higher order, thinking humans, are also necessarily materialists. They search for truth in matter because there is nowhere else for them to search: all they can see, hear and feel is matter.

—Anton Chekhov, letter to Alexei Suvorin, May 7, 1889

CONTENTS

INTRODUCTION

JUST THREE SYSTEMS

Now for the materialls themselves, I reduce them unto two sorts; one Naturall, of which some are more familiarly known & named amongst us, as divers sorts of Birds, foure-footed Beasts and Fishes.... shell-Creatures, Insects, Mineralls, Outlandish-Fruits, and the like.... The other sort is Artificialls, as Utensills, Householdstuffe, Habits, Instruments of Warre used by severall Nations, rare curiosities of Art, &c.

—John Tradescant, from his catalogue to the *Musaeum Tradescantianum* (1656)[1]

THERE ARE SEVEN SPECIES of great ape on the planet. Six of them live in nature. One cannot live without artificial aid. Humans would die without tools, clothes, fire, and shelter. So how, if technology compensates us for everything we do, did we ever manage to evolve in the first place? With such innate deficits, how did the weakest ape come out on top?

This is the story of our remarkable ascent. I am an archaeologist and prehistorian, and a devout agnostic, disturbed by the aggressive illogicality and fraud of creationism but also unhappy with the conventional biological account of human evolution. Mostly I believe in the world that we see and touch, the one we are born into. Before we can speak or hold onto a thought, our senses of sight, touch, taste, and smell drink in the material reality that lies all around us. Before we know it—literally—the world we are born into becomes internalized. This is how the world now is, irrespective of what is natural and given, and what is artificial and culturally created. And perhaps because of this profound and primal familiarity, I believe we have continually underestimated the effect of this physical reality in making us human.

This book traces humanity back more than 2 million years, long before writing and cities, before the use of metals or farming, to a point just after we diverged from the ancestors of chimpanzees, gorillas, and orangutans (two species of each of which make up the other six great apes). And it does not avoid the central question of our precise origin.

Religion can demand its moment of creation, yet it has no data. Science envisions a long gray scale along which no firm line between apes as apes, apes as ancestors, and ancestors as the earliest humans can ever be drawn. This book insists that there was an actual moment when we became human. It was a moment long before we became intelligent in any modern sense. It was a moment seized by a female as, for the very first time, she turned to technology to protect her child. In that moment, everything that we were going to become was made not just possible but inevitable.

Archaeology is about recovering the remains of the distant past, bringing back to light what has fallen out of mind, piecing together lives perhaps abruptly cut short, but long since over anyway. Sometimes we recover treasure, sometimes trash. And, alongside artifacts and debris are the physical remains of the people who made them and used them—fragments of bone.

It is midsummer in the English countryside, in the region of North Yorkshire, so it is no surprise that wet hail rakes Ingleborough Hill. Underfoot the limestone has been eaten out by millennia of rain, the water feeding fast-forming rivulets that sink into numberless underground fissures, clefts, pots, undiscovered caves, squeezes, and pits. At Gaping Gill, a powerful stream falls out of sight through a vast subterranean cavern large enough to hold the great medieval cathedral of York Minster. We are entering one of a myriad of tiny sinkholes, known to cavers as Y-Pot.

Having scrambled down to the choke—a rough platform made of jammed-in limestone blocks—I am supposed to post myself sideways

FIGURE 1 Fragments of a child's skeleton, provisionally dated to the Bronze Age, from Y-Pot, Ingleborough, North Yorkshire. (Photo courtesy the author.)

into a narrow slot. It hardly approaches body width, and my caving harness will not fit. Attached to a simple waist rope, I slide backward, free my wedged helmet, and find a metal rung inside the top of the shaft. John Thorp, known by his cave-tag, Lugger, and a veteran of archaeological cave exploration, is just below while his fellow caver Deb Limbert belays the rope down from above and supplies me with necessary words of encouragement. I pay minute attention, as instructed, to Lugger's descending body position, half off the ladder, right hand on rungs, left arm pinned back by a rising fin of limestone. I follow to where the shaft opens sharply out and the metal ladder drops freely. Standing on the unstable surface of a lower choke, beneath which lie more voids, my glasses steam up. But I can just make out the place where, in black mud, Lugger has found the skull of a child.

Did it fall or was it pushed? Were the bones washed in to become mixed with those of red deer? The black mud under our feet is full of bones. The analysis of the animal species, examination of the child skull (spongy-looking bone inside the top of the eye-sockets indicated chronic anemia), and then the extraction of bone collagen suitable for radiocarbon dating, are part of the basics of modern archaeology. But right now I am thinking more about how I get back out. The passage in was so tight that I had had to remove the flashlight from my top pocket to get through; going back up, slick with wet brown silt, will be a birth against gravity, though perhaps not as tricky or as risky as an ordinary human birth.

Human beings should not exist. Our skulls are so large that we risk being stuck and dying even as we are struggling to be born. Helped out by a technical team—obstetrician, midwife, and a battery of bleeping machines—the unwieldy cranium is followed into the light by a pathetic excuse for a mammalian body, screaming, hairless, and so muscularly feeble that it has no chance of supporting its head properly for months. How did a species in which basic reproduction is so easily fatal, and whose progeny need several years of adult support before they can dress themselves, not just evolve but become the dominant species on its planet, capable of inhabiting virtually every environment, from the tropics to the poles, the coasts to the high continental plateaux, prospecting under mountains and across the ocean floor?

In becoming so dominant we have had to contend with many forces, from gravity, pressure, and temperature, to predators and pathogens, and on to ourselves and the consequences of our own technology in creating pollution and conflict alongside insulated comfort. These forces belong to three systems, and only three. Each has a different logic. The first is the system of physics and nonbiological chemistry, where fundamental forces create the elementary particles and energy quanta, hold them together, and regularize their interaction. This system extends from the very small scale of nuclear physics and below to the very large scale of cosmic star formation. It has governed our planet from the time of its formation, before there was any life, and it continues to this day with its fundamental rules unalterable. The gravity acting on my body in the cave is part of this system.

The second system is that of biology, the algebra of life. This system operates at a medium scale, mostly on the outermost layer of our planet, on the land and in the oceans. We do not yet know whether it occurs anywhere but on our planet. Probably it does, but proof awaits. This second system has to conform to the first: the bodies of animals and plants must respect gravity, contend with solar radiation, keep the diffusion and osmosis of elements across cell walls in order. They must do all this by respecting the rules of chemistry and physics. But they embody the higher logic of genetics. Life is built of special molecules with complex qualities and emergent properties. The most critical of these is replication. The unfolding of replication, and the incorporation of random errors (mutations) in successor generations of replicated complex molecules, lies at the heart of the process of evolution. This system was first uncovered by Charles Darwin, and like the first system of inanimate nature, living nature is a focus of immense human investigation. But humans bring us to the third great system, and the most disputed and controversial. If the first two systems are natural, one nonliving, the other living, then the third is artificial.

The physical environment in the cave is clear enough: the exchange of heat and moisture in the cave space, the greater proximity of the base of the shaft to the earth's center, and the ineluctable action of gravity. The biological interweaves with the physical here, as the limestone rocks around us were once the living bodies of corals. More recent deaths are represented by the bones of deer and other vertebrate animals, and by those of the child. These remains are made of chemicals—carbon, calcium, oxygen, nitrogen, sulfur, strontium, and so on—that were taken up and given form according to genetic instructions. Now they have been released back into smaller life systems, becoming reabsorbed at the microbial and submicrobial levels, eventually to become fossilized back into the rock that is itself made of the compressed exoskeletons of billions of dead coral polyps. For a brief time, these deer, boar, and wolves battled entropy, exerted energy to do those simple things that can never happen in the nonliving system, such as running uphill against the direction of gravity.

The third system is represented by our ropes and clothing, helmets and LED headlamps, cameras, plastic lunch boxes, watches, and glasses. Although in this list the things all belong to the modern material world, the child down here, an individual who lived, we estimate, in the Bronze Age, also once had a world of technology supporting him or her. The clues are there, around us, awaiting archaeological interpretation: the red deer bones scattered in the cave were animals whose pelts were used for clothing; the scattered woodland in the valleys was used for fuel. The underground world we have descended into began to be prospected around 4,000 years ago for blue and green ores that could be burned, heated, transformed, purified, and made into what we now call metal. This child would have known and seen how shapes and forms could be constructed from natural materials, bringing previously unknown objects, such as bronze axes and picks, into existence.

Our relationship with material products is bizarre and complex but also straightforward and familiar. We use things not just to adapt to the natural world, manipulate the laws of the physical, and subvert the instincts of the biological: we use them to construct ourselves.

Item: A.D. 1626: English naturalist John Tradescant displays both "Naturalls" and "Artificialls" in his Cabinet of Curiosities, the world's first pay-per-view museum; it includes the hand of a mermaid and a splinter from the True Cross.

Item: A.D. 2008: Widowed by the destruction of the Berlin Wall, a woman marries a garden fence.

Item: 650,000 years ago: Someone is scalped, their whole skull defleshed, perhaps for food, leaving seventeen chipped stone tool marks on their skull.

Item: A.D. 2006: In eye makeup and black head-to-toe garb, a terrorist suspect uses his sister's passport to evade capture at airport border control.

Item: A.D. 1792: On a beach in Tasmania, a tribe demands that a shipload of French sailors remove their trousers to expose the configuration of their sexual anatomy.

Item: 25,000 years ago: A chunk of limestone is carved into a naked human figure, with explicit genitalia but no face, then coated in red ocher.

Item: A.D. 2007: eBay agent 2513 pays $15,269.69 in a charity auction for two used breast implants.

Item: A.D. 1943: A poet observes a soldier through a telescopic sight, reflects that his target has a mother who loves him, and kills him.

Item: A.D. 1320: The owner of a pair of scissors and a lancet is flattened by a pagan standing stone he hopes to see demolished.

Item: 2800 B.C.: 18 million hours of shared effort over two years results in a new, synthetic hill in a landscape of natural ones.

Item: 2.6 million years ago: In a sheltered spot out of the wild, some human ancestors settle down to a day shaping rocks with other rocks to make tools to tilt the balance of the natural and the biological decisively in their favor.

Item: A.D. 2010: A three-month-old girl dies of malnutrition because her parents are too busy caring for a virtual child at an Internet café.

Item: 3.5 million years ago: An upright walking female ape, struggling to hip-carry a tiny infant, begins to fall behind the rest of her group.[2]

We use, believe in, take for granted, become obsessed by, die for, and kill with things. Not just any things, not natural things. What is most distinctively human about us is our relationship with artifice. Imagine a soccer player in a world without soccer balls, a seamstress without needles, a child without a blanket, a reader without a book. Human life as we know it assumes the presence of artifice—objects we have made ourselves, without which life would either have no meaning or be physically impossible. Not only did we make these necessary objects, but, within a framework of some 2 or 3 million years, the objects have physically and mentally shaped us. Without them or their incursion into our lives, our heads would be a different size, our body type would be different, we wouldn't be living in houses. There would be no houses.

I will attempt to persuade you that Darwin was wrong, not because I believe in alien intervention, whether from UFOs or from God. Not because I believe the weak-minded claims of so-called intelligent design. But because the very tools Darwin used to understand the natural world—from HMS *Beagle* to Down House, from pens and paper to microscopes and scalpels—subvert biology, break the rules of evolution, and undermine natural selection.

Technology is a word that comes from the ancient Greek *technē*, meaning a craft, the art of doing something, which required technique and theory to achieve.[3] Without the fabricated objects of technology, our artifacts, inventions, gizmos, consumer items, manufactures (whatever words we choose for them), not only would we seem less than human, but we could never have actually evolved. Our relationship with technology and material culture is the critical point of distinction between us and animals.

This book does not deny that human evolution has happened, but it argues that although Darwin was correct that we evolved from simpler creatures, he was fundamentally wrong about its causes and the conventional "reasons" given for our evolution. The idea that we, with our weak eyes, fragile backs, and infantile helplessness, are the result of the brutal sifting of a blindly operating process of natural selection is as hard to make sense of as the competing creationist idea that we were made in all our naked inadequacy by God (or, at least, somehow "intelligently designed"). I believe both explanations are wrong—although the creationists are more wrong—and for precisely the same reason: they disregard the role that artificial aids played in our development, and the fact that without them, we are utterly hopeless. Yet with them, we are our planet's dominant force.

I want to take us back to the beginning. Or, at least to a beginning, at a point when humans started walking upright. Walking upright (bipedalism) meant that hands became free, with an amazing new potential. All that was then needed was an intelligence powerful enough to direct the hands to make things. I believe that from the very start our early ancestors took control over their own evolution by developing technologies: from baby slings to furs, from hunting spears to flint, from fires to cave hearths and other shelters. As our ranks grew, technology developed into levers and pulleys and carts and implements to cultivate land. We did not have to be the strongest or the fastest because we now had machines doing the heavy lifting. These eventually included cars, wheelchairs, elevators, Jet Skis, and aircraft, so that we no longer needed functional hind legs. Televisions, mobile phones, and the Internet meant we no longer needed to physically travel anyway. In the modern industrialized world, technology has grown so powerful that assembly-line robots and computer-controlled cybernetic systems can fabricate things without us. By organizing our society along the lines of a complex machine, with function-specific parts, some of us have been wholly freed from the daily routines previously essential for personal maintenance. Food, transport, clothing, shelter, warmth, light—all can be provided to allow specialists in technological research and development to focus full-time on analyzing the microscopic inner-level functioning of materials. So the materials at our disposal are no longer limited to those discovered in the natural world, like stones. Nor do they stop at the basic transformations accomplished by natural processes, such as heating ore to produce metal. We no longer just manipulate and recombine materials, but invent new materials, from scratch, modeling their molecular form within computers and assembling them, atom by atom, using nanotechnology.

There is increasing evidence that we are no longer governed by natural selection. Technology can and does supersede biology and lead us into a new form of life, one not primarily governed by Darwinian process. The implications of being the first entity on our planet to escape natural selection are immense. We have never been wholly natural creatures, and we have evolved to be increasingly artificial. Even should we want it, escape from technology is no longer possible. It may in fact be that technology has escaped us: the inertia of the entire system of technological civilization is by now so immense that the sorts of choices left for us to make in the future are essentially trivial. The ride we are now on may be unsustainable, or it may not; but there are many reasons for believing that we are incapable of getting off. Either we crash, or we continue our artificial ascent. There is no soft landing into a quieter and more balanced world that utopian souls often dream of.

I am sure that extraordinary opportunities lie ahead, and that to take them we need to know what has happened. We need to finally understand who we are as humans, and where we came from. We must stop living the lie that we are either animals or divine creations. In this book I will argue not only that we must learn to love the machine because it represents our only viable future, but that, for far longer than we imagine, we have been part of the machine. We long ago began adapting our minds and bodies to a hidden agenda. The result is a symbiotic form of life—one that breaks the old rules, and whose new rules (if they exist) are far from understood.

Paradoxically, and most controversially, I want to claim that we did not somehow naturally become smart enough to invent the technology on which we critically rely and that has removed us from the effects of natural selection. Instead, the technology evolved us.

I aim to demonstrate by clear examples that rather than humans evolving to become intelligent enough to invent tools and weapons, shelters, monuments, art, and writing, the objects, in the most critical instances, came first. Our changing biological capacities, physical and mental, positive and negative, followed. Our innate qualities have been altered beyond recognition away from the simple processes of survival of the fittest. We have long been absolutely unable to survive without the artificial realm. It insulates us, cures us, compensates for our deficiencies of sight, mobility, metabolism, and memory. The realm of technology may not have a mind of its own, in the sense that we understand our individual, personal existence, but it has a powerful unfolding logic that may ultimately amount to the same thing.

The idea of a third system of powers and laws is already accepted within artificial intelligence theory. One of the leading researchers, Ray Kurzweil, enthralled by the ability of computers to out-think mere humans (especially in terms of mathematical calculation and applied problems involving complex strategy within particular well-defined settings, such as the game of chess), looks forward to a more fundamental moment of overhaul, when we humans get comprehensively left behind. He calls that point "the singularity."[4] It is the point when computers will have become more intelligent than humans across all the most important ranges of intellectual endeavor. In a similar vein (but with a significant difference of emphasis), Kevin Kelly has identified an essentially autonomous world of technology, a realm he calls "the technium."

Kelly says that he tends to think of the technium "like a child of humanity," and his vision presupposes a fairly clear biology-technology divide.[5] Kurzweil's singularity concept also considers an imminent and dramatic sundering. But the distinctions may not exist for humans in quite the way we have hitherto imagined. We are talking more than just synergy here: our evolution, uniquely, is the history of elision between biological substrate and artificial construct. I share Kelly and Kurzweil's belief that something unprecedented and nonbiological is under way, but I see it a little differently and set it loose far, far earlier. My System 3 emerges almost coterminously with human biological origins, but significantly predating and underwriting them.[6] This means that as a species we belong in part to both System 2 and System 3 (in a similar way to viruses, which straddle Systems 1 and 2). As the philosopher John Gray has provocatively noted, "Descartes described animals as machines. The great cogitator would have been nearer the truth if he had described himself as a machine. Consciousness may be the human attribute that machines can most easily reproduce. It may be in their capacity for consciousness that humans and the machines they are now devising are most alike."[7]

"We" are not limited by biology but represent a kind of symbiosis, a way in which intelligence extends back out into the inanimate and deathless. There is no need to fear a cold, robotic vision—things will not replace us, anymore than we can do without them. It is the division that is unreal.

FIGURE 2 Patterning in three systems: System 1, beach-washed natural pebble (left); System 2, fossil mollusk (center); System 3, fragment from flint tool production (right). (Photo courtesy the author.)

It might yet all go wrong. Many times in the past, a loss of control over technology, or a misunderstanding of its broader environmental implications, has caused a civilization to collapse. Society breaks down, useful science is attacked, dangerous superstitions grow. Although, eventually, all has been rebuilt and we have reached a new, and higher, level of technological expertise, there is no guarantee that we can endlessly make the same mistakes. We require a necessary momentum to build the future. For us, there is no nature we can go back to anymore. The only way out is up.

Humans are not bounded by the biological body in the same way that animals are. We are extended through artifacts, and they, in turn, are extended through human biology. It is not even correct to say that artifacts reverberate through "us," as if they were different and separable from us. In critical ways we are them and they are us. Why else would one die to defend a flag, risk death to dive for pearls, or be so involved in scientific or artistic works that one overlooks the apparent imperative of biological reproduction? Those phenomena do not necessarily represent sacrifices or dysfunction; they are typical of humans, and they clearly confound any simple Darwinian logic in the same way that a deer walking uphill confounds the logic of gravity.

This book sets the nature-nurture debate on one side and enters newer, less familiar territory. It juxtaposes past and present, comparing the world's earliest puppet, carved out of mammoth ivory some 26,000 years ago, and the body of the contemporary artist Antony Gormley multiplied in life-size iron casts along a beach;[8] contrasts the simple technologies of the modern humans who first spread around the world before and during the last Ice Age with the apparently even simpler technology of the Aboriginal Tasmanians; and considers the significance of the extraordinary prehistoric organic artifacts and structures that have survived in the Hallstatt salt mines in Austria in relation to the austere record of durable stone that is typically all that remains at the vast majority of prehistoric findspots.

Starting from the way human beings turn evolution inside out, we trace the limits of the Darwinian explanation of our species' existence, behavior, appearance, and attributes. Through a reassessment of the technological and cultural status of the Aboriginal Tasmanians, the idea of "natural" humanity is called into question, and key questions posed about the order in which things really happened during our prehistoric evolution. This is then used to throw a critical and surprising light on what is currently happening to our bodies and minds, why they are progressively and inevitably weakening, and why it may not ultimately matter.

I argue that although we are still powerfully biological, and often constrained by the evolutionary processes Darwin described, we are clearly more. We remain constrained by gravity in many ways, but this does not mean that we are mere physical entities. Birds use flight to overcome this physical force—biology overcoming basic nonbiological physics. Humans use artificial technology to understand this physical force and the flight of birds, and then travel beyond the domain of terrestrial systems altogether. Although space flight was far in the future, the remains we found in the cave shaft were those of an artificial ape.

CHAPTER 1

SURVIVAL OF THE WEAKEST

A fox (canis fulvipes), of a kind said to be peculiar to the island, and very rare in it, and which is a new species, was sitting on the rocks. He was so intently absorbed in watching the work of the officers, that I was able, by quietly walking up behind, to knock him on the head with my geological hammer.
—Charles Darwin, *Journal of Researches*, December 6, 1834[1]

IN THE HIGH ALPS a 5,000-year-old frozen corpse, "Ötzi the Ice Man," lay beside his longbow. The axe marks had not been smoothed off, but the functional aspects were complete—it was a replacement weapon, made in haste. We know he was a man in a hurry: sophisticated forensic archaeological analysis has revealed his story, from the sequence of inhaled tree pollens in his lungs, characteristic of different altitudes, to his stomach contents and the scans that show a rib broken only days before death. He had, my Austrian colleagues are now convinced, been on the run, up mountains and down valleys, for several days, living rough, pursued, desperate to rearm himself, and finally driven toward a high mountain pass with a blizzard threatening. There he was shot in the back from a distance by an expert. Still running, he reached back to grasp the shaft where it projected beneath his shoulder blade; it snapped, leaving the flint arrowhead lodged. The internal bleeding was unstoppable, and by now out of sight of his enemies, he collapsed into the first heavy snow of winter and died. Ice closed over him, and he vanished from memory until 1991, when global warming, acting on the Ötztal glacier, melted him back into view.[2] By the time of Ötzi's death, in the later Neolithic period around 3300 B.C., human beings had long been the planet's top predator, fearing mostly each other.

Before Ötzi died, he had been a farmer who, from time to time, also hunted. He was part of a wave of colonization and economic change that had been sweeping into central Europe from the Balkans (and ultimately western Asia) since 5500 B.C. His ancestors had carved out territory in a zone previously dominated by dedicated wilderness dwellers. These were people who had never had domestic animals or fenced off land. They were mobile and innovative bands of men, women, and children who hunted, fished, and gathered. Their communities had expanded rapidly into the burgeoning new forests of Europe around 10,000 years ago, after the ice sheets receded and the climate quickly warmed. These Mesolithic people were in turn descended from big-game hunters whose lineage can be traced back to a time when mammoth-hunting scenes were painted on cave walls.[3]

This period, the Upper Paleolithic, was when the previous lords of Europe, the Neanderthals, were being squeezed out by our kind, anatomically modern *Homo sapiens*, first arriving from the Near East around 40,000 years ago. Neanderthals had weathered successive periods of advancing and retreating ice over the previous quarter of a million years. Before them had been *Homo heidelbergensis*, and, yet further back, *Homo erectus*.[4] The sequence of kinds of humans, and kinds of human culture, arriving in waves over the hundreds of millennia before Ötzi's death is at least as complex as the history of Europe since that date. To understand how Ötzi (and we) came to look physically modern, to be able to think intelligently and make use of tools and weapons, we will have to track back far earlier, to a point even before there were humans.

I am unconvinced by the current understanding of our evolution—not whether or when evolutionary changes took place, but how and why. Like my academic colleagues, I have no doubt that humans emerged over vast spans of time, that our distant ancestors were apes, and that we are biological creatures. The results of each new field season of research in paleoanthropology—a new pelvis from Ethiopia,[5] a different sort of small-headed *Homo erectus* from the high Caucasus

Mountains,[6] a new, very early tree-living ape that might be placed on or off our direct ancestral line according to highly technical arguments and scholarly taste[7]—important as they are, do not change the basic fact that humans evolved from apes.

This simple fact is often misunderstood: "If we evolved from chimpanzees, then why are there still chimpanzees?" asked one child, after an introductory lecture I had given his school class, sure that he had spotted the killer weakness in the evolution argument. The answer, of course, is that we didn't. Chimpanzees have evolved alongside us since a point, maybe 7 million years ago, when we formed a single breeding community of some kind of ape. That ape, whatever it was, was much more like a chimpanzee than like us, and that means that modern chimpanzees have not evolved as dramatically as we have. One group of apes continued doing what they had been doing pretty successfully. They may have moved down from trees more, and evolved knuckle-walking, but they never walked fully upright.[8] They kept their insulating fur, never finding a need for clothes, never freeing up their hands to make them, or developing brains large enough to think of them. The other group, split off perhaps by a simple fact of geography, got embroiled in something else. What happened to them was very different and very varied: before we emerged as the only surviving upright-walking ape, there were numerous prototypes, most of them evolutionary dead ends.

Darwin had no access to the African fossil record—he suspected it might exist, but he did not live long enough to see even the first finds. But he did apply some smart thinking to the living species of ape he knew about. He considered it significant that two kinds of great ape—gorillas and chimpanzees (two species of each are now recognized)—lived in the wild in Africa while only one—the orangutan (also now divided into two species)—lived wild outside it, in southeast Asia. Using the same logic he had applied to the evolution of birds and fish, Darwin argued that Africa, as home to the most kinds of great ape, should be the continent on which the common ancestor of apes and humans had lived. He predicted, in advance of the fossil evidence, that it would be there that any intermediate "missing links" would be discovered.

Darwin upheld his African origin theory even after the discovery of the first Neanderthal skeleton in Germany, believing (correctly, as it turns out) that Neanderthals were relatively modern, being a parallel line of later human evolution and not an intermediate link with apes. It was nearly a decade after Darwin's death that the first physical evidence for a missing link came to light, at Trinil in Java. Although "Java Man," now classified as *Homo erectus*, was, like the orangutan, also outside Africa, paleontological fieldwork throughout the twentieth century and up to the present shows that the most significant phases of human evolution happened in Africa.[9]

Discoveries from the African Rift Valley, down into southern Africa and across to northern Africa—everywhere that conditions have been conducive to the fossilized preservation of ancient bone—follow so thick and fast that it is hard to publish a detailed scheme of human evolution without its being immediately out of date. The evolutionary tree (or bush) of human and humanlike evolution is continually added to and reformatted, with all the controversy that entails. Several plausible versions exist at any one time as scholars connect the dots between a bewildering number of different upright-walking ape species that existed, often in parallel, between 5 and 1 million years ago. Upright-walking apes are collectively known as hominins; in the light of genetic studies, the older and more familiar term, hominid, has been extended to include chimpanzees, gorillas, and orangutans. All of this—the analysis and reanalysis, the dot-joining, and the terminological evolution—is science, and the myth-bound creationists who claim that the revisions, occasional mistakes, and even the rare misrepresentations of paleoanthropologists and archaeologists betray uncertainty about whether evolution happened fail to comprehend that evidence-based research, at its best, has no dogma.

Every year, new discoveries add complexity to the several million years' worth of explosive changes that differentiate our species from hairy, knuckle-walking, inarticulate apes. Chimpanzee, gorilla, and orangutan intelligence has been maligned in the past; their potential for communication using sounds and tokens, along with their tool-using abilities and level of social complexity, was previously underrated; and respectable estimates of genetic closeness put them within 5 percent of "being human." But this book is not about how *like us* they are, but the mystery of how *different* we have managed to become, and continue to become. Biological logic alone cannot explain it.

Biological evolution is both fact and theory. Fact in the way that the earth's roundness is fact: our world may look and feel flat, but we can sail or fly across it without changing direction and get back to where we started. Equally, we can now watch biological evolution in the lab as new strains of virus and bacterium develop, at times too rapidly for adequate medical response. This is something we are desperate

FIGURE 3 A schematic graphic of the basics of human evolution (to be read clockwise): A reconstruction of *Sahelanthropus tchadensis* is shown at 2 o'clock; it may not actually be the common ancestor species of both chimpanzees (1 o'clock) and modern humans (11 o'clock), but it is currently one of the closest candidates. From this bifurcation point, gracile australipithecines arise (3 o'clock) and from them robust australopithecines (also known as paranthropines, 4 o'clock). These go extinct, but early genus *Homo* also evolves from the australopiths (5 o'clock: the reconstruction can be read as *H. habilis* or *H. rudolfensis*). Whether *Homo ergaster* (7 o'clock) arises from the habilines or from some as yet unknown line of descent from gracile austropiths is not known for certain. From *H. ergaster* comes *H. erectus* (8 o'clock), which eventually goes extinct, as well as the line that leads to *Homo heidelbergensis* (9 o'clock), which many see as the last common ancestor of Neanderthals (10 o'clock) and moderns. (Graphic © Frankland/Taylor.)

to understand better, and our struggle gives birth to theory, which is tested and refined. In the same way, theory supplies scenarios for the condensation of stars and the birth of solar systems like ours, in which spherical planets can form. The best current theory for the creation of sphericality remains uncertain and incomplete (gravity is involved, and we do not completely understand that yet). The condensation of our planet only happened once, so we cannot go back and check it. But that does not make the world flat: we cannot sail off the edge of the ocean into some medieval vision of Hell.

Even so, the ability to view our blue planet floating as a globe in space has brought home the reality; and human evolution, while no less real, has no such indisputable graphic proof. Nevertheless, the basic facts are clear: we descended from apes, lost our fur, began to walk upright, trebled or quadrupled our brain capacity, and created a culture that includes tools, language, ideas, and values, ultimately becoming smart enough to look back over this process and analyze it.

Ötzi, like us and like his earlier prehistoric ancestors, did not face the world naked. He had elaborate gear—for protection, for making, for killing. When he took his bow into the forest to hunt deer, he did not do what all other predators do—lurk, chase, and harry until an old, weak animal is brought to bay, or a young fawn is separated from its mother. Ötzi would have left the old and weak for lesser predators like wolf and lynx, and would have started the fawn in the right direction to rejoin its family. Ötzi had an interest in managing the wild. He was after the biggest and best—it would provide the most meat, and the most status when its antlers were placed above his cabin door.

Because he had a bow, Ötzi undermined the logic of evolution. In the old biological arms race, predators who were too slow starved, and the fastest prey escaped, with the result that the fittest in each species survived to pass on its genes. But the idea of individual fitness driving evolution is an approximation. More precisely, every species has an "inclusive fitness" that represents the sum total reproductive effectiveness of its genes in a particular environment. Some extremely fit animals may sacrifice themselves to protect close kin, or so that the broader group lives on in good numbers, preserving a better complement of the species' "best" genes. Species also tweak each other's inclusive fitness. In North America, wolves, by hunting elk (moose), increase their prey's inclusive fitness by removing the old and diseased from the breeding pool. And just by chasing their prey around, wolf packs limit the time the elk can stay in large herds in one place—conditions under which diseases would otherwise easily become endemic.[10] The wolves keep the elk fit, and vice versa: wolf packs with lower average stamina or less social cohesion will do less well, catching fewer elk and producing fewer offspring to reproduce themselves.

Once humans emerge, this biological arms race is replaced by a more familiar (and real) one. Tools and weapons have been called the "extra-somatic means of adaptation for the human organism,"[11] enhancing innate somatic (or body) strength beyond what would seem naturally possible (this complex of artifacts and technological know-how that belong to my System 3).The unprecedented power of a longbow like Ötzi's allows even a hunter with a serious physical disadvantage (bad legs, say) to ambush and bring down the fittest beast in a herd. Through technology, the laws of nature are supplanted by the will of humans.

Darwin's little fox, living quietly on an island off the coast of Chile, had no response to the hammer that the great naturalist wielded, and ended up, stuffed, in the collection of the Royal Zoological Society in London. But the real weapon was not the hammer but the ship, the *Beagle*—a floating support system that allowed a contingent of our species to arrive unannounced almost anywhere in the world. During the *Beagle*'s voyage the body count rose swiftly. The crew killed to eat—as much from curiosity as hunger (Darwin ate a puma in Patagonia)—and sometimes to drink (as when Darwin slashed open the bladder of a live Australian land tortoise during a desert crossing). They also killed to protect themselves, but principally they killed to provide the basis for the most important zoological study of all time. The voyage of the *Beagle* led to the development of the magnificent theory of evolution by natural selection, and the idea of the survival of the fittest. So it is ironic that the specimens themselves were selected artificially, with the weak, small, and ill rejected in favor of those that, until their untimely deaths, had been in peak condition.

Darwin's ideas dominated intellectual conversation in late Victorian Britain. While the novelist George Eliot wrote in one of her letters, "Natural selection is not always good and depends (see Darwin) on many caprices of very foolish animals," capturing very honestly the bleak logic of natural selection, others wanted a teleological explanation.[12] That is, they wanted nature to have an *aim*. If it could no longer be God-given beauty, at least it could be fitness—an idea of perfect adaptation to a given environment. The phrase "survival of the fittest" was coined not by Darwin but by a follower with a more social and political agenda, Herbert Spencer.[13] Nevertheless, Darwin was persuaded to adopt the phrase after the fourth printing of *The Origin of Species by Means of Natural Selection*.[14] As an enthusiastic hunter, he must have been keenly aware that, while the slowest antelope would be eaten by the cheetah, an antelope fit enough to outrun a cheetah was thereby a prize trophy, potentially bringing praise, status, and perhaps even enhanced access to reproduction for its human killer. So which would be the "fitter"?

One fitness ploy for antelopes might be to develop superb camouflage, outwitting both cheetahs and humans. But such coloration might look so dull that the antelope would not find a mate. Darwin recognized that animals had, at times, to strut their stuff, whatever the risk. Without an overarching, single-strand idea of fitness, analyzing the pressures that caused species to evolve was very challenging, and Darwin developed his theory of sexual selection to take care of part of the problem. The classic exposition of sexual selection relates to the peacock's tail (and has become infamous because of the way Darwin extrapolated from it to explain why—in his opinion—his wife, Emma Wedgwood, was less smart than he was).[15]

The clash between different kinds of fitness, and between the mechanisms of natural selection (external to a species as it battles for survival) and sexual selection (internal to a species as its members consider potential mates), puts tension into biological systems. That this creates unpredictable results is well known. Nor is the directed interference of humans in the evolutionary paths of other animals unfamiliar. Most environmental scientists, genetic engineers, and evolutionary biologists agree that nature is increasingly under the direct control of science and technology. We have seized the reins of evolution, and even if we cannot direct it with full confidence, we are changing its course. Genetic engineering is the obvious modern example, but if we look at all the domesticated farm animals and pets around us, it becomes clear that our active interference has a long history. The natural equations of wild survival are in recession: in future, we alone will define fitness.

Although the power of technology today is unprecedented, the tipping point occurred over 2.5 million years ago. The dawn of the technological era is signaled archaeologically by the first chipped stone artifact—a tool or weapon plausibly used for killing big game. After that point, for animals confronted by humans, the characteristics that would ordinarily convey fitness could increasingly become a liability. The process of natural selection and survival of the fittest was undermined. Intelligent humans with weapons could kill whichever animals they liked, fit or unfit, young or old, large or small, and the animals, trapped by the biology of inheritance, had no effective response.

Some species of super-heavy, protein-laden herbivores, such as the mammoth in the Old World and the mastodon in the New World, were doomed: too meaty, too heavy, too slow to adapt, they rapidly went extinct. But among medium-size herbivores, such as the forest cattle and boar, mountain goats and wild horses, the smallest and least aggressive were often spared by humans. While their prize-specimen relatives were trophy-hunted and butchered for the pot, they found themselves penned. Cattle decreased dramatically in size: wild aurochs bulls (*Bos primigenius*) weigh in, like American and European bison, at around 2,200 pounds. After domestication kicked in, from around 7,500 years ago at sites like Catalhöyük in Anatolia, Mehrgahr in the Indus Valley, and Nabta Playa in Egypt, domestic species emerged that were half that size or smaller. *Bos taurus*, developed in the Near East, became the base species of most modern cattle because it was traded so widely. By 4000 B.C. it had been adopted throughout continental Europe to reach Ireland in the west and China and Korea in the east.[16] En route, the animals were discerningly bred in every generation, creating the characteristics that would make them docile, easy to transport and overwinter indoors, and productive of milk and meat. Apart from those bred specifically as brawny draft oxen, the trend was ever smaller. In the northern Isles of Scotland, descendants of these first Neolithic cattle include the tiny Norland breed, whose fully grown bulls weigh a mere 440 pounds, with females just 176 pounds.[17]

Such animals, which would normally have lost out to the larger, more dominant beasts, found themselves captive, their opportunities defined for them. They were to provide meat on demand, whenever a hunter—now an early livestock farmer—needed it. By Ötzi's time, 3300 B.C., it was not just meat, but the milk that animals produced for their own young that was taken, along with their labor. Ridden, yoked, laden with baggage, or bred to get fat fast—a series of key species were so changed by domestication that nowadays most medium-size land animals cannot survive in the wild. They depend on humans for their maintenance, food, protection, health, the terms of their death, and their chances of reproduction.

Darwin used his understanding of the activities of domestic animal breeders to provide a framework for what happened in nature. He knew the pedigrees of many of the more familiar farmyard breeds of cow, sheep, and pig, studied form among race horses, conversed with dog breeders, and joined clubs of racing pigeon enthusiasts. Among this fraternity there was a powerful idea of "bloodlines," and of pure, mixed, and degenerate blood. It may have been unscientific, but with no genetic science to help, it was the best way of imagining what was going on. Darwin's cousin, Francis Galton, keen to uncover the physical basis by which particular qualities were passed down the generations to be enhanced or eventually fade out, had a household full of variegated rabbit colonies: white, black, brown, gray. Among these unfortunate creatures, he swapped blood by transfusion, in the (vain) hope that the coats of their offspring might change color. Galton was not working with the correct concepts; unknown to him, to Darwin, and to the rest of the scientific world, the foundations of the idea of heritability, and therefore of genetics, were being uncovered by one Gregor Mendel, working among his sweet peas and bees in a sleepy Augustinian monastery in Brno. Nevertheless, Galton was on the hunt for a way to directly convert the mechanisms of selection seen in nature into controlled artifice, so that new kinds of animal and plant might be artificially brought into existence.[18]

The intense interest in artificial selection for desirable characteristics had a formative effect on Darwin. Though the mechanism remained obscure, he saw that once species became domesticated, they were endlessly mutated by people. Over the space of a few generations, wild cattle had been turned into not just beef herds, milch cows, and plowoxen, but a plethora of functionally specific breeds, suitable for different environments. From a single ancestral sheep stock came Welsh mountain sheep that could jump across mountain streams and little ravines, and shaggy Swaledales that could not jump and stayed enclosed by the drystone walls of the otherwise open limestone landscape of the Yorkshire Dales. The British National Sheep Association lists eighty-six breeds, and many other parts of the world have their own range, but there are even more types of horse. These include drays for pulling great carts, shire horses for plowing, hardy quarter horses for the human and baggage carrying that won the Old West, heavy cavalry steeds, show jumpers, foxhunters, flat racers, children's ponies....

In fact, the ornamental peacocks whose tails fascinated Darwin belonged to garden breeds of wild peafowl (*Pavo cristatus*), domesticated as early as 1000 B.C. in Egypt, and probably much earlier in China. It turns out that we cannot tell how much of the peacock's plumage is natural, how much was sexually selected in the wild by peahens (despite its potential liability), and how much elaborated artificially in the walled pleasure gardens of Xanadu. If humans had effected so much variation, then, Darwin reasoned, natural processes, working over millions of years—a vast scale inferred by the new geologists of the previous generation—had had ample time to make greater changes.

As volcanoes produced new land, sea beds were raised, and ocean levels fluctuated to split islands apart, with the result that the populations of individual species became isolated. Because various species faced different challenges, the terms of fitness were subtly, or blatantly, altered for the separate "founder" groups. Though they shared a common ancestor, one population of a species might develop white fur for camouflage in snow, while another population, no longer in contact, remained brown. Over immense spans of time, life, which began in the seas, came onto land. Early amphibians were followed by reptiles and then dinosaurs, who were later ousted by birds and mammals. Each major type had its generalist survivors and its niche specialists, its run-of-the-mill members and its eccentric outliers. Some successful body-plans remained very stable over time, as in the insects known as roachids, which emerged 300 million years ago in the Carboniferous era, and from which the modern kinds of cockroach evolved 200 million years later, in the Cretaceous. In other cases, evolution effects dramatic and rapid change. Through the study of DNA, relationships that seem outwardly implausible are being uncovered, as between the hippopotamus and the dolphin, which both descend from a common, mud-wallowing ancestor. The cetacean branch returned as mammals to an ocean that millions of years previously saw their emergence as lungfish. Given enough time, fins became legs, which again became fins (though with a different internal architecture: the whale, as Herman Melville fondly asserted in *Moby Dick*, is indeed a kind of fish).

What is peculiarly difficult to grasp about human evolution (compared to that of any other animal) is that it is both biological and cultural. Recalling that there are three basic systems in operation in the universe, it is clear that everything on earth is equally constrained by the inanimate, physical, and chemical rules of nature. Animals and plants also conform to the rules of the second system, mostly connected with classic Darwinian evolution, but with other processes (such as the direct swapping of genetic information between simple life forms) and the complexities of evolutionary development (gene switching) also in operation. The complexities need not detain us here. The key point is that humans are only partially bound by this second system, having used cultural knowledge to instantiate the third, technological realm.

There is another way of considering this argument. A number of researchers, particularly biologists, think of human culture as an elaboration on biology, a sort of artificial peacock tail that must, ultimately, obey the same rules of reproduction and survival as everything else in the living world. "Meme theory" is one of the terms applied to this sort of thinking.[19] In it, human technology is seen as little different from birds' nests, hermit crab shells, or any other material prop that a species uses for its adaptation in a particular natural environment. Because we can analyze a bird's nest or a spider's web as having architectural qualities, we can easily trick ourselves into thinking that their architecture and our architecture are the same kind of thing, constrained by a similar universal Darwinian logic of evolutionary success and failure, so that culture becomes a (not very neat) subset of broader biology. I will explain later why we must reject this.

If we accept, for the moment, that humans belong to biological systems but are also profoundly affected by and dependent on technocultural systems, which may well have an independent logic, then there are several different routes our development can take. The pure biological pathway is, thanks to Darwin and the discovery of DNA-based genes, understandable in terms of central principles. If a larger lung capacity and shorter, smaller, warmth-retaining extremities convey an advantage in the struggle to live at high altitude, humans with those attributes are more likely to survive and breed. Over time, this will give rise to a particular physical type in high-altitude regions. This is what we observe among both Himalayan and Andean populations: short and stocky people, with short fingers and barrel chests. As the anatomist and paleoanthropologist Christopher Ruff, of Johns Hopkins University, has demonstrated in a study of the average bodily dimensions of seventy-one human groups,[20] our species conforms closely to two well-known zoological rules: populations in the colder zones, typically nearer to the poles, have a stocky build, with low surface area to body mass, maximizing heat retention; by contrast, populations in hotter climes near the equator, regardless of whether they are tall or short, have limbs that are thinner and longer relative to their bodies, maximizing heat loss (these are known as Bergmann's Rule and Allen's Rule).[21]

The purely material or technological aspects of human cultural production such as tools, art, and clothing embody the principles of the third system. With artificial objects, the processes involved are more complex, lacking the dependable algebra of the biological realm. But we still know a good deal about the invention of stone arrowheads, the wheel, metals, and writing. We can see how innovations spread, and how they change the way people live, without affecting their chances of biological reproduction in any obvious way.

Where things become radically complicated is when biological and technological systems interact, in one direction or another, sometimes both at once. In these circumstances, we have to grasp the feedback between genetics, nutrition, and natural environment, and the alterations that the advent of System 3 both allowed and caused. Because biological evolution was there before culture, we often give it precedence. The effect of biology on the production of familiar items of material culture is clear. Without a certain level of intellect and a certain kind of manual dexterity, we could not have constructed either the sophisticated mammoth-bone, stone, and skin houses found at Molodova 1 in Ukraine, dating to 44,000 years ago, nor the rather larger, 36-story habitation platform known as the Saturn V rocket—probably the most complex artifact yet made—that successfully launched our species beyond the dominance of our local planetary gravity.[22] The same skills are manifested in everything in between: pottery night-lights (from 30,000 years ago), the discovery of the uses of metal (from 10,000 years ago), standard systems of weights and measures, the invention of the wheel, and writing (all between 4000 and 3000 B.C.),[23] or—my

FIGURE 4 Bronze Age wooden staircase, prehistoric salt mine, Hallstatt, Austria. (Photo: anwora © NHM Vienna.)

personal favorite—the breathtaking modular-assembly mine staircase with alterable tread angle, recently discovered in the prehistoric salt mine of Hallstatt in Austria, its timbers felled, according to dendrochronology, in the years 1344 and 1343 B.C. (figure 4).[24]

But some of the changes detectable at these times show our techno-cultural world impinging significantly on biology, in a process termed gene-culture co-evolution. Although the domestication of cattle potentially allowed the spread of dairy farming, the very earliest *Bos taurus* in the Near East were meat-only animals. The idea of using the nutrient supply intended for infants of another species for one's own sustenance probably had to overcome psychological and deep-rooted traditional objections. Anyone who tried it really would have felt weird, as the adult human back then had no way to digest such a product. Milk sugar, lactase, can be broken down only by the enzyme lactose, and that is, or used to be, produced only in the stomachs of newborns.

But someone must have seen the advantages of giving children animal milk. In the new domestic environment it would not have been surprising. In contrast to the preceding hunter-gatherer period, the farming revolution tipped adult women into ever more of a routine daily grind, quite literally. At the early Neolithic site of Tel Abu Hureyra, the women, to judge from their skeletons, spent much of each day on all fours, feet and ankles braced, hands grasping a loaf-shaped grit-stone and grinding flour in a saddle quern (while the men were either tending herds or hunting). The cereal meal so arduously produced could maintain dietary levels year-round, and when simmered in new-fangled pottery vessels, the gruel could be given to babies. In this way their mothers were "freed" to keep grinding.[25]

In the less than pretty powerhouse of Near Eastern farming, women increasingly found themselves on all fours. When not making flour, they produced children. The effect of interruption of the long-evolved, on-demand hunter-gatherer pattern of breastfeeding with all its natural contraceptive benefits was dramatic.[26] The natural four-year gap between children was halved, and although infant mortality rose too, the population began to burgeon.

It was the British Prime Minister and war leader Winston Churchill who, with his characteristically direct understanding of the history of human competitiveness, said, "there is no finer investment for any community than putting milk into babies."[27] Once animal milk had been found suitable for human children, the adults must have begun to think about how such a useful animal protein, extractable while the animal's meat remained intact and alive, could be enhanced. Presumably, children who retained the ability to digest lactose as they grew older found themselves at enough of an advantage when starvation and famine struck early Neolithic communities that their genes were favored. What we do know is that, increasingly after around 8,000 years ago, widespread genetic change among both Eurasian and African populations allowed lactase synthesis into adulthood.

As we shall see, the way we absorb food more generally, especially our need for highly processed, high-energy, high-protein foods to power large brains perched on inadequately short lengths of gut (an unfortunate side effect of the switch to upright walking), is based on the knowledge of cooking, fermenting, and curing through which crucial extra caloric value can be gained.[28]

But if key aspects of our biology would be impossibly dysfunctional without technological support, how did they first evolve? Looked at one way, the pivotal biological change of growing the smarter brains allowed cultural inventions like big-game hunting with spears and making fire to cook meat. Looked at the other way, these same cultural inventions are what allowed us to possess larger brains (and shorter guts) in the first place. The solution to this puzzle must in some way connect with the undeniable fact that humans, at a very early stage, turned the survival—and therefore evolution—of the creatures around them upside down. The fit became targets, the weak survived to be valued as "fit" in terms of which we are the sole arbiters.

To understand this we have to look much further back than the little mammoth-tent village of Molodova, to a time before the Upper Paleolithic with its populations of essentially biologically modern humans. Moving deeper into the remote past, we face a puzzle. How do we know what mental capacities existed before modern humans existed? As we track back past *Homo sapiens* and Neanderthals to *Homo erectus* and then to other genera, australopithecines and paranthropines, from whom or next to whom we originally evolved, how can we judge from the combinations of preserved bones and surviving artifacts what, precisely, was going on, and what the chain of causality was?

Even the apparently less problematic last 10,000 years, following the global rise of farming, calls our fundamental biology into question. If we have such a profound effect on the wild around us, bringing it inside our evolving domestic sphere, rendering it docile and subservient, what might we have done to ourselves? There is a hint that we too have become domestic animals, selected for all the features of bodily weakness. Where are the great ripping canine teeth of our chimpanzee and gorilla cousins? The hard, tearing claws? Where is our thick pelt of insulating fur, our intimidating muscle power? We may have our strong men and women, Olympians and pugilists, but, unarmed, they are no match for a gorilla. Yet we are clearly now the dominant species. Could it be that the emergence of technology took the strain, and allowed us (through subprocesses that Darwin would have understood well enough) to evolve biological deficits? Having possession of fire, tools, weapons, and clothes, we do not need massive teeth, claws, and muscles, or a long, vegetable-absorbing gut. Here is the central question of this book: Did we evolve into the modern human form, and then invent the objects on which we now depend, or did the things come first and so bring us into being? If it is the latter (as I believe), then we are first and foremost the product not of natural selection or sexual selection but of a special kind of artificial selection.

Biologically speaking, "fitness" is the ability to adapt to one's environment and reproduce, just passing on one's genes. But I shall use terms such as de-evolution and survival of the weakest to provoke us to think about the special way in which we are "biologically reduced." Even in the last 10,000 years (the blink of an eye in evolutionary time), our bodies have weakened dramatically. Over this timescale it can be shown that our stature has decreased by 7 percent: Christopher Ruff estimates that we have lost fully 10 percent of our overall bony ruggedness—our so-called skeletal robusticity—in that time.[29] Over the past 100,000 years, we see a 30 percent overall decrease—not as great as in some of the cattle we have domesticated, but remarkable nonetheless. Even Özti, living just 5,000 years ago, had significantly stronger bones than most of us.[30]

There is an energetic logic to this. Because growing and maintaining a large, robust skeleton is costly in energy terms, allowing emergent technology to take the strain makes good sense. Of course, that was not a decision that archaic humans made, or would have been able to make, consciously, but it was an inevitable biological consequence of creating a wider range of tools to do jobs that previously relied solely on muscle power. A more gracile body will need less upkeep, and what it can no longer manage by brute force can be managed with specifically designed artifacts that amplify and concentrate strength: slingshots and spears, levers and bows. These technologies allowed our self-domestication just as they aided our domestication of wild animals.

The human skeleton is not static, but responds to activity, strengthening itself when the muscles are used more, and becoming more fragile when a person is habitually inactive. These same patterns are discerned in teeth, yet they are much more directly controlled by genes. This implies that a powerful evolutionary process is at work, with identifiable causes, yet there is a great deal of disagreement about what precisely is going on, and why.

I am interested in the causes of our unique intelligence. While our bodies became significantly weaker, following our evolutionary split from the apelike ancestor we share with gorillas and chimps, this one feature dramatically increased in power. Human intelligence provides the underpinning for making bows and arrows, rifles, rockets, satellite tracking systems...not to mention the quieter but no less critical achievements of weaving, basketry, and ceramic manufacture. Most scientific accounts of our species' unprecedented ascent agree that the key question is *why* we became so smart. Darwin's theory of sexual selection purported to answer this but was based on questionable assumptions about the relations between the sexes in the past, coupled with an anecdotal and prejudiced assessment of the relative intelligence of women and men in the present. Turning things around, this book asks not why, but how we gained intelligence. I want to know *how was it that we did not remain stupid* when evolutionary pressure should have made brain expansion beyond the chimpanzee level physically possible.[31]

Let us return to Ötzi, a few weeks before his murder by another human, out game hunting. What do we observe, looking at things with a broadly biological eye? Obviously a creature on its hind legs. But then, it *has* only hind legs, as what were once, long, long ago, forelegs have evolved into arms. That is unusual, but not wholly unique. On the savannahs of Africa, and in the rainforests there, as in Borneo, there are apes with forelimbs that we can call arms; and there are monkeys in South America that swing through the branches using arms and legs in different, mutually supporting ways. What is unique in our Neolithic Ice Man is the complete uprightness, the habitual, unsupported vertical gait. Ötzi could crawl (if he wanted to), or swing from a branch, but he did not need to do either once he had learned to walk.

Once Ötzi had also learned to talk and use his hands, he could copy, take instruction, and, eventually, work out how to think for himself. By then he was a fearful predator and opponent. He could impale and shoot, set traps and snares, dealing death at a distance, in either space or time. The things he used to accomplish these feats—net, spear, bow—were things he knew how to make and repair, personalize, and (perhaps) even improve. He had everything with him—a rucksack with a bentwood frame, a fire-making toolkit, a pouch with medical remedies, a stone axe. He even had one of the much-sought-after, newfangled copper axes. These, in the central Europe of 3300 B.C., were too valuable even to serve as ceremonial grave goods, instead being passed down to the next generation.

Ötzi's toolkit existed in tandem with the contents of his cranium which, as compared to an ape's, had grown three- or fourfold. I have already suggested that humans break the external rules of natural selection. They interfere in nature in counter-biological ways, targeting prey in a novel fashion, deploying technology to subvert the natural logic of survival that had hitherto always existed to drive evolution a particular way. Ötzi, like the mystery child in Y-Pot, was an artificial ape but not, perhaps, quite as artificially protected as we are now. In most prehistoric societies it seems likely that a harsh fate awaited the physically weak and the disabled. Although Ötzi had a remarkably sophisticated set of gear that could extend his powers, he still needed swift reflexes, high levels of stamina, and sharp eyesight if he was to stay ahead of the game. There were no telescopic sights, no corrective spectacles, no radar tracking. If you were good enough when you drew your bow, then what you saw was what you got. If not, you went home hungry.

In the modern West, over 60 percent of the population is thought to require the assistance of an optician to enjoy optimal sight. Of course, some of this is just the price we pay for our longevity. Failing sight resulting from old age would have been less noticeable in Stone Age populations, when relatively few people lived beyond their forties. But in modern America, 10 percent of preschool children have measurable visual impairment, and the percentage rises to 25 percent by the time children are eleven or twelve years old.[32] We assume that this does not matter much as optometrists are smart enough to have the artificial aids to correct the deficiency (which they often do). But what if the development of optical technology has removed the selection pressure on naturally good eyesight? There is evidence that this could well be the case.

Darwin, when the *Beagle* brought him into contact with several of the tribes of Tierra del Fuego at the southernmost tip of South America, was shocked by the rigors of the climate and the inhabitants' lack of protection against it ("a woman . . . suckling a recently-born child . . . whilst the sleet fell and thawed on her naked bosom, and on the skin of her naked baby!"[33]), yet was impressed by the superior natural senses of the tribespeople. Three Fuegians were already on board the *Beagle* when it left England: York Minster, Fuegia Basket, and Jemmy Button (the last so named because Captain Robert Fitzroy had bought him as a little boy from his parents for a price of one pearl button). In his journal Darwin observed: "Their sight was remarkably acute: it is well known that sailors from long practice, can make out a distant object much better than a landsman; but both York and Jemmy were much superior to any sailor on board: several times they have declared what some distant object has been, and though doubted by every one, they have proved right, when it has been examined through a telescope. They were quite conscious of this power; and Jemmy, when he had any little quarrel with the officer on watch, would say, 'Me see ship, me no tell.'" Darwin was not given to hyperbole, and by stating that the Fuegians' sight was "*much* superior" to that of "*any* sailor on board"[34] (my emphasis), he was drawing attention to a matter of great interest.

Darwin painted a picture of an extreme battle for survival among the Fuegians. We shall later turn to a now-controversial account of the old being cannibalized in time of famine.[35] Whether or not that happened (and I believe it did), the overall timbre of Darwin's account of this culture is of unremitting, ruthless competition, even within families. The Fuegians inhabiting this freezing southern tip of the Americas existed under precisely the conditions where the selection pressures that could weed out anyone with substandard sight, hearing, mobility, or comprehension operated with maximum force. Children who could not learn to do things well and fast stood little chance of reaching adulthood and passing their genes on.

The Fuegians lived what appeared to be one of the most abject existences that Darwin ever encountered. He thought their lives were miserable in the extreme. But physically they were very powerful, retaining the most rugged and robust skeletons known among the grades of modern humanity. Elsewhere, a more elaborated culture has insulated our species. Most of the peoples of the world, especially in the urban centers that have arisen with the past 5,000 years, are spared many, and often most, immediate and acute survival pressures.

Yet our culture may finally have become too clever at protecting and providing for us. Not only may our natural capacities, like brute strength and visual acuity, be weakening, but our brains too may, after a long period of evolutionary expansion, be at last growing smaller.

Researchers like Peter Brown calculate that our brains shrank on average by 9.5 percent in the last 10,000 years alone (more, as a proportion, than our bodies, for which he recorded the 7 percent drop already mentioned).[36] Further back, over 100,000 years ago, we find that the Neanderthals had, on average, bigger brains than ours. We would like to think that is because they had bigger bodies, too—a simple proportional effect. Either that or their brains were less efficient, like the big gas-guzzling engines of old cars as compared to the high-performance fuel-sippers of the present. Both things could be true. But we should also consider the possibility that, with fewer cultural props around them, Neanderthals had to be naturally smarter, had to think for themselves more. As in Tierra del Fuego, passengers could rarely be carried. Perhaps our victory over them will eventually have a hollow ring. The third system, the realm of artifacts, may even now be producing the ultimate reverb on our innate biological capacities.

We may call the strange evolutionary flow the "survival of the weakest" or a type of de-evolution. Of course, there is a rhetoric in this, as there was to the slogan Spencer coined from Darwin's ideas, "survival of the fittest." But I want to point out what may be illogical about the Darwinian account of human nature. It has to do with the measurement of success.

Now we shall embark on a much longer journey, spanning the deep past, the present, and the near future. It is about how the world of manufactured artifacts—tools, weapons, technology, machines—has changed, and continues to change, the underlying terms of evolution. The process that led to the development of projectile-launching devices, like Ötzi's bow, short-circuited the rules of fitness in the animals it targeted, allowing what had once been the fittest to be hunted and the weaker ones to become domesticated and so live on. Slowly, we too lost our physicality, farming it out to the increasingly essential toolkit of things we now need simply to exist. All the while our brains were able to expand, via remarkable and accidental feedback mechanisms.

Chapter 2

NAKED CUNNING

I could not have believed how wide was the difference between savage and civilised man: it is greater than between a wild and domesticated animal.

—Charles Darwin, *The Origin of Species*[1]

IF OUR USE OF TECHNOLOGY has altered our physical and mental evolution, how is it that some people seem to have lived in harmony with the natural world, not needing to subjugate it technologically? Surely their existence disproves my case and supports the more familiar idea that it is only recently that, for good or ill, we have become "artificial." Especially now that primatologists are discovering more and more about chimpanzee and gorilla tool use, including the fact that different ape communities pass down different technique traditions, the boundary between the simplest human technologies and the most complex ape ones seems to have broken down.

The Aboriginal Tasmanians stand out as a people whose stone tools appeared as simple as those used by chimps. Many of them demonstrably were. If the Tasmanians lived as natural people in wild nature, then my concept of the artificial ape—a being that has, uniquely, crossed a critical threshold—will make no sense. My thesis runs against Darwin's conception of a slow gradation of biological evolution linking humans to apes, a line along which the Aborigines could be, and were, placed: to the left, earlier, closer to apes. Today, it is conventional to place those gradations in a distant past, all of it before about half a million years ago, a period populated by a plethora of missing links. But in Darwin's day the evidence was considered to be alive and kicking, made manifest not just in the different outward appearances of different peoples and races, but also in the grades of their intelligence as betrayed by an unequal capacity for developing technological civilization.

Anchored off the cold, rain-swept coast of Tierra del Fuego in the summer of 1832–1833, Darwin wrote, "I shall never forget how wild and savage one group appeared: suddenly four or five men came to the edge of an overhanging cliff; they were absolutely naked, and their long hair streamed about their faces; they held rugged staffs in their hands, and, springing from the ground, they waved their arms round their heads, and sent forth the most hideous yells." Darwin was unsettled by the idea that these "stunted, miserable wretches," whose "language... according to our notions, scarcely deserves to be called articulate," were actually human.[2] Yet the Tasmanians appeared to be, if anything, more bestial.

At the time of European contact, the Tasmanians lived an exposed, frequently roofless life in a climate where one might least expect it. Paul de Strzelecki, a Polish aristocrat, geologist, and man of letters, visiting a couple of years before Darwin, wrote that Tasmania was "cold, wet, and liable to sudden changes of temperature, where bathing ceases to be a pleasure."[3] Exposed to such rigors, the naked skin of the natives—for they wore no clothes—appeared "scaly, spotted by... disease, and weather-beaten." The people were houseless wanderers, sleeping in the open or behind temporary windbreaks. Incredibly, they were apparently unable to make fire, having always to carry it with them, as fire sticks or as embers inside hollowed wood fire logs. When it rained and the fire went out, the group were left cold until they could locate a forest fire (not always easy) or run across some other band who had escaped inundation and get a spark back. Even sworn enemies could not refuse such a critical request.

When the *Beagle*, with a young Darwin aboard, anchored in Hobart in 1836, the last indigenous Tasmanian community had recently been deported to a tiny offshore island. He never met them, but his impressions of the mainland New South Wales Aborigines (who were so far in advance of the Tasmanians that they could make fire) were nevertheless negative. He grudgingly recognized a form of humanity and acknowledged the operation of certain colonial prejudices. But although they were "far from being such utterly degraded beings as they have usually been represented," Darwin reckoned them lower in the "scale of civilization." He wrote: "In their own arts they are admirable...they will not, however, cultivate the ground, or build houses and remain stationary, or even take the trouble of tending a flock of sheep when given to them."[4]

His mature reflection, expressed in 1871 in *The Descent of Man*, was that "the civilised races of man will almost certainly exterminate, and replace, the savage races throughout the world."[5] The gradation he perceived as linking "the negro or Australian and the gorilla" with the baboon at the lower end and "the ever-improving white or Caucasian race at the upper end" would be broken, and the traces of our human evolution from apes would become even less apparent: "the break between man and his nearest allies will then be wider." Darwin graded the Aboriginal Australians on the bottom rung of his evolutionary scheme. Ranked below Africans and Fuegians, they were "the lowest barbarians"; it followed that the Aboriginal subset that had just been deported from Tasmania was in last place among all the regional peoples of the antipodes and as close to our species' evolutionary origins as it was possible to be. They were the creatures against whom human progress could be objectively gauged.

Darwin's near contemporary, the German ethnologist Gustav Klemm, viewed such people as *Naturvolk*, belonging to nature and lacking the vital spark of culture. *Kulturvölker*—people of culture—such as the Germans and British, but also the ancient Greeks, Chinese, Egyptians, Aztec, and other civilization builders, were considered not just to be more culturally elaborate, but to belong to a group innately capable of cultural innovation.[6] By contrast, the *Naturvolk* were not deemed capable of invention. They could consider themselves lucky to become the passive recipients of pots or wheels, writing, sails, or formal architecture, the precious items of progress. Klemm's assessment marked a reaction against the optimistic view of human nature that had dominated European thought in the previous century, embodied in the French philosopher Jean-Jacques Rousseau's idea of the noble savage, all humans being essentially "children of nature."

Native Tasmanians first definitively encountered Europeans in 1772, when the French, under Marc-Joseph Marion du Fresne, made landfall, and an unpleasant misunderstanding occurred.[7] Du Fresne was a child of the Enlightenment, convinced by idealists at home, such as Rousseau, that the human family was all one under the skin. Not everyone agreed, but there was not yet a Darwin to dignify counterclaims of racial inequality with a solid scientific theory. Rousseau thought culture an exterior acquisition, valuable, but not inbred. Progress and betterment were potentially available to all. By removing your clothes, discarding the insignia of rank and class, you could become one with the "noble savages" of far-flung islands, and du Fresne was determined that, whenever natives were sighted, his men, at least, should meet them on an equal footing. Anchoring in what later became known as Blackman's Bay, he ordered two sailors to row ashore naked.

We will later examine why this particular encounter ended in tragedy (as gifts were exchanged and not fully comprehended, by either side). But, even if it had not, in terms of broader geopolitics the French were being squeezed out by the British. In 1770, James Cook had claimed Australia for George III, and Tasmania was to follow. The years 1802 to 1807 witnessed the first buildings going up in Hobart. Alongside estate managers, mineral prospectors, and émigré visionaries, there were many convicts. Paul de Strzelecki considered these white colonists "the outcasts of society... accustomed to treat with contempt any rights which their brutal strength could bear down"; they "invaded the natives' hunting-grounds, seized their women, and gave rise to [a] frightful system of bloody attacks and reprisals."[8]

Darwinian evolutionary theory provided the scientific rationale that allowed the Aboriginal Tasmanian nation to be labeled as "the most degraded and brutal in the world," "the lowest known species of the human family, just a step higher than the chimpanzee."[9] They were classified as hopelessly and innately stupid, and they fell victim to abduction, rape, murder, enslavement, and transportation, as well as alcoholism and syphilis. The last indigenous pre-contact Tasmanian, Truganini, "a savage maiden, trained in the wilderness," died in miserable circumstances in 1876.[10] With her death, the official extinction of the Tasmanian Aboriginal race was trumpeted by the colonial Tasmanian government, and the passing of this last of the "typical representatives of the Early Stone and Wood Age"[11] was hailed as a "distinct step in human progress."[12]

What is hard to grasp even now is that the colonizers of Tasmania were not ignorant of other cultures. The ships of the Dutch, the French, and the British had polyglot crews, and had put into port all over the world. They were not, in any straightforward way, narrow-minded. But the Tasmanians were a peculiar provocation, living perversely illogical lives. Stark naked in the bitter cold, not only could they not make fire, but they also made no fishhooks. Although they were surrounded by a teeming ocean, fish were taboo; when abducted as slaves for the settlers or by the Malay whalers, the Aborigines resisted eating anything with scales or fins. Left to their own devices, they beachcombed, looking for crabs and mussels, or for seals to club. Greased with seal fat, the women dived for abalone and lobster in the icy waters.

Things appeared little better inland. Tasmanian settlements were hardly worthy of the name. Some tribes made thatched huts, but most seemed to use only roofless windbreaks. Like mainland Australian Aborigines, they had no writing, no draft animals, wheels, bows, arrows, or pottery, and—despite the island's rich deposits of silver, lead, tin, iron, copper, and gold—no knowledge of metal. Unlike mainland Aborigines, they had no bone tools, hafted axes, ground stone tools, boomerangs, or nets. Without clothes, they used grease not only when diving, but also to protect themselves from the frequent heavy rains. Mixing the grease with clay and red ocher, they put it in their hair, "a prey to filthiness... in order to prevent the generation of vermin."[13] The tools they had were made from phthanite, a fine-grained sandstone, which, unlike flint, produces no spark when knapped.

The European ships' crews, in the starkest contrast, were adept in all varieties of tool use, manufacture, and repair, from the simple and elegant to the complicated and confounding. They knew how to maintain, adapt, and invent, negotiating a network of things—a multifaceted technological "material culture"—to insulate themselves against whatever nature, or other humans, threw at them. This was what had allowed them to push the boundaries of the possible, afloat for years at a time, and become the first circumnavigators of the planet.

Many had sailed among Caribs, Fijians, and Andaman Islanders; others were familiar with the Inuit and Aleut. The better-traveled among them could have observed that the tropical cultures had plenty of stuff—fine hunting and fishing equipment, ruthless arsenals of weapons, strong traditions of house- and boat-building—but that much of it was ornate and symbolic indulgence. The groups of the far north on the other hand, living in hostile waters, hunting elusive and dangerous prey, had fewer frills but even more extraordinary gear: fantastic kayaks, a wide range of different types of nets, harpoons, bows, spears, traps, and clothing sewn from marine mammal pelts, from all grades of feather and down, and even the taut waterproof skin of birds' feet.

It was obvious that the Tasmanians had none of this. Sometimes they would adorn their bodies with charcoal or red ocher, but without hereditary chiefs, there was no regalia of office (tribal elders, headmen, and spirit mediums were recognized, but Truganini, for example, was dubbed a "queen" only with a patronizing colonial wink). They had no discernible marriage ceremonies, and a loose polygamy prevailed. Inheritance was what you learned from your elders.

The Tasmanians were non-literate, while many of the sailors and settlers were illiterate. The crucial difference here lay in the perception of deficits and the discernment of a drive to betterment or its absence. The whites knew what to aspire to, even if they fell short. The Tasmanians, arrogant in their ignorance, let humanity down. The whites, feeling keen shame for their educational and moral inadequacies, were liberated in the presence of a whole population of inferiors, officially validated as such by the world's leading scholars. They could bully and abuse, write off the tribespeople, and deny them basic rights and charity. On the native side, a 10,000-year isolation had created a mindset not only resistant to change from outside, but wholly unaware until it actually materialized that there was an outside from which change could come.[14]

It is accepted that what unfolded after European settlement was an atrocity, but that is usually where modern thinking stops. How the Tasmanian lifestyle came about—how its cultural losses made sense—is something we no longer inquire into. The "Tasmanian story" has become less about how the tribes existed before contact, and more about the evils of racism and how not to treat other cultures.

Fire making was brought out of Africa as a key skill in the progressive colonization of the Old and New Worlds by our distant ancestors. This was the first great human diaspora, across Eurasia, down into Australia, and across the Bering Land Bridge into the Americas. The journey had many stages and lasted many tens of thousands of years. The great age of European discovery, from Marco Polo and Columbus through to Captain Cook and Semonov,[15] was no more than a reconnecting of the dots, bringing a species, diversified into almost every ecological niche, back into contact with itself. The first humans to reach Australia definitely brought fire making with them. Geologically, Tasmania is a southern continuation of the mountains of Australia's eastern flank. When ice ages lock up vast quantities of seawater, lowlands emerge to connect Tasmania to Australia, morphing the island into a peninsula. The last time this happened, 37,000 years ago, humans had just started to occupy Australia, arriving by a series of short ocean hops from New Guinea. They needed boats for the initial continental colonization, but not to reach Tasmania. To do that, one only had to walk. And 35,000 years ago, they did, as mobile hunter-gatherer bands, in a landscape that included the previously submerged lowlands. As more and more water was locked into ice, the shorelines extended ever farther from the peak of Mount Ossa.[16]

The mountain grew in stature as the sea level sank, and the highlands became bleaker. In an increasingly glacial world, the fertile coastal margin became the best place to be. Although the traces of the most populous occupation have long been covered by returning ocean, enough archaeological evidence survives from coastal caves (then far inland) to show that the earliest Tasmanians weathered the last Ice Age in style, hunting, gathering, and fishing. They were clothed too: remains of bone needles and stone tools for cleaning hides suggest that they (like their Ice Age counterparts in Europe) used leather and fur, sewn with sinew, to insulate themselves from the elements. And, like the rest of the population of the continent, they certainly used fire.

Then, 10,000 years ago, meltwaters from the Southern Ocean glaciers began to cut Tasmania off. Within a few hundred years, a population of perhaps 5,000 had become cut off from mainland Australia in a rather definitive way. Unlike the Polynesians, whose culture centered on sophisticated outrigger canoes with which the scattered archipelagos of the warmer and more predictable Pacific had been colonized, the Tasmanians were pedestrian colonists isolated by rapid topographic and climatic flux. Yet, even with good boats they would have been isolated practically speaking—the ripping tides, chains of small islands, and semisubmerged rocks of Bass Strait are largely shunned by sailing vessels even today. In reality, the little rafts that the Tasmanians sometimes built at the time of European contact could stay afloat only for a few miles even in fair weather.

The Tasmanian relationship with the ocean may have worsened progressively after the island was cut off. Archaeological sites that date back over 5,000 years—cave shelters like Cape South Cave and Rocky Cape—do not seem to show any fish-eating taboo, being stuffed with fish bones and worked-bone points. These points were not directly used to catch fish because they were not barbed, and no fishhooks have been found. So how did they get the fish? Some archaeologists think the points were used for making fishing nets, others that they were for hanging bait inside box traps. As the Aborigines at the time of the first European contact had neither nets nor traps, perhaps they were just basic spear points. Between 4,000 and 3,500 years ago, these objects disappeared, and after that, there were no more bone tools.

As the ancient technology analyst Bill McGrew has shown, the Tasmanians went from tools with several elements—"techno-units"—to tools with just one.[17] Ötzi the Ice Man's gear was all multipart: a yew haft for a copper axe head, held in place with resins and leather thong, the bentwood and leather rucksack, a stone point set into a wooden armature to sharpen his goose-feather-fletched arrows. By contrast, all-wood spears, where the tip is simply fire hardened, along with stone scrapers and hand choppers, are single-element tools. The multiple-element (composite) tools are typical of almost all human groups studied by anthropologists and prehistorians; the single-element tools, as McGrew argued in a controversial comparison, are typical only of chimpanzees…and the last Tasmanians.

These Aboriginal Tasmanians, lacking axes to fell trees to hollow out canoes or build permanent houses, lacking bone awls and needles to cobble shoes or sew clothing, have recently taken center stage in discussion of how a culture might go downhill without its individual members being congenitally stupid (that is, somehow biologically "less evolved"). In his influential 1997 account of the overall shape of human history, *Guns, Germs, and Steel*, the geographer and anthropologist Jared Diamond blames the situation of the Aboriginal tribes not on an innate deficit but on accidents of history and geography.[18] A powerful advocate for a nonracist account of human differences, he argues that the Tasmanians simply became isolated, and so were peculiarly vulnerable to the vagaries of time and chance. Diamond weighs up three possible explanations for the paucity of Tasmanian material culture compared with that of mainland Australian Aborigines. Either the things the Tasmanians lacked had been introduced to the continent from southeast Asia relatively recently, and were therefore unable to spread south before Tasmania was cut off, or the new things had been invented in mainland Australia after Bass Strait had formed. Diamond argued that only after ruling out these two possibilities should we consider the third: that the Tasmanians arrived with a suite of inventions, shared with their mainland cousins, but after they were cut off, they somehow conspired to lose them.

Hunting with dingoes, for instance, had been practiced by the mainland Aborigines only since these semi-wild dogs were brought in from southeast Asia around 1500 B.C., long after Tasmania was cut off. Similarly, the techniques for manufacturing microliths—tiny, elegant stone tools that can be mounted on wooden armatures using bitumen to make light arrows, drills, and other composite tools—seem to have been mastered on the mainland only after the land bridge disappeared. But more basic stone tools, shaped and sharpened by grinding, had been present in Australia at least 15,000 years before Tasmania was cut off, and the Tasmanians lacked even these. The boomerang is also at least 10,000 years old (probably far older) but absent from Tasmania.

Bone points, however, provably fell into disuse in Tasmania, something Jared Diamond judges "a significant loss, because warm clothing sewn with bone needles would surely have been useful in Tasmanian winters."[19] He then turns to the fish question: "remains at archaeological sites show that Tasmanians used to catch many fish species, which accounted for about ten percent of their calorie intake. Most of the species they caught are still common and easy to catch in Tasmanian waters today." This sets up his key questions: "Do societies really do such maladaptive things? And if so, why?" He turns to the work of archaeologist Rhys Jones for an answer.

In 1978, Jones, following excavation of a series of key sites and the development of new, more accurate methods of radiocarbon dating, wrote with conviction: "There in its stark simplicity of about two dozen items is the entire corpus of Tasmanian technology. No simpler technology has ever been recorded in the world's ethnographic literature."[20] "In the closed system of the Tasmanians, maladaptation might have a better chance of surviving simply because of a lack of better-competing neighboring communities. If fish were not caught for several generations, the isolated Tasmanians would have had no opportunity of relearning such skills from neighbors even if they had wanted to."[21] Diamond agrees, but he does not concede Jones's further conclusion, that the Tasmanians suffered a "squeezing of intellectuality"; or that, through their technological losses, they were doomed to "a slow strangulation of the mind."[22] More sympathetically, Diamond believes that the Tasmanians were unable to "repent their folly" when first some, then all, fish species became taboo, and then the art of making a particular kind of bone tool was lost. He supposes it far easier for techniques to disappear over time, essentially accidentally, than for new ones to be invented. With no commerce beyond their island, Tasmanians were cut adrift from the pulse of renewal or change.

But neither Jones nor Diamond gets near a plausible explanation of why practices they both regard as commonsensical in the Tasmanian environment should be abandoned or tabooed. The fish mystery, along with the fire-making mystery, has provoked some bizarre solutions. One postapocalyptic vision (reminiscent of William Golding's *Lord of the Flies*)[23] encourages us to imagine a few plucky children surviving a massive outbreak of toxic fish poisoning caused by a "red tide" of marine dinoflagelles, an algal bloom phenomenon more common in warmer waters. The children's parents, who have eaten more fish, all die, along with grandparents and infants. Thus the secrets of fire making and bone-tool carving are lost. Tabooing the sea as evil, the children have to start society up again with only a basic toolkit.

This scenario stretches plausibility in many ways. Children in small tribal societies learn the techniques for making tools and fire early in life, out of necessity, and they can hardly be imagined to be old enough to survive alone, ignorant of such basics. Yet the outlandish hypothesis provides a useful reminder of just what we have to explain, how hard it is to make sense of. Under what circumstances could fire making—a skill passed down from parent to child in an unbroken 2-million-year survival chain—go missing?

The story of *Guns, Germs, and Steel* is rooted in a style of thought older than Victorian racism—the "geographical possibilism" of the eighteenth century. This tried to explain differences in cultural patterns, not in terms of differences in human racial capacities, but in terms of what particular environments will allow and what they rule out. Often now called ecological determinism, it is a sort of universal acid, creating "just-so" stories of why people live as they do in particular environments. This way of thinking is fine when people in similar types of places are discovered to live similar lives, nomadically wandering over great plains wherever great plains are to be found, or settling in permanent villages next to fish-rich estuaries when the opportunity presents itself. But it is challenged by the Tasmanians, who "should" be equipped more like the native peoples of the high-latitude north. Why should the high-latitude south produce an almost opposite culture?

Without matches or some other easy technology, I cannot make fire. I even remember forgetting how to make it, not long after I had learned. I was taught at an ancient technology fair at Grand Pressigny, a French town famed for its prehistoric flint quarries. I was presented with a strike-a-light kit comprising flint, iron striker, and two types of tinder: seed-pod material (like cotton wool) and a short length of rope with a frayed end. After a couple of hours of tutorial, wine at elbow, I was able to strike sparks into the cotton tinder, light the frayed rope, and blow on it to keep a glow going and light any number of fires. On returning home, I found myself unable to reproduce my success for my daughters. Initially curious, they quickly became bored in a way typical of humans seeking enlightenment from a patent incompetent.

The kit remains in a corner of my office, attracting dust and thoughts of inadequacy. By briefly learning the art of fire making and not practicing enough to ingrain it as second nature, I reflect the post-manufacturing, post-industrial culture in which I live. The benefits of fire are effortlessly there when I switch on the microwave, turn up the heat, or run a bath. I could consult one of the many "how-to" websites (such as primitivefire.com), but at present there is no pressing need to know (and when there is, the website may not be there).

Diamond belongs to a broad consensus of academics who believe that Darwin got it wrong about the Aborigines, although he does not say it outright: in the face of rising intellectual panic over religious fundamentalism, directly criticizing the father of evolution may be seen as irresponsible, and Darwin's views on race are usually passed over lightly in the hope that they can be discreetly decoupled from his overarching conception of human evolution.

If Darwin was wrong that the Tasmanians represented an earlier, more primitive stage of biological evolution, is Diamond right that their deficiencies arose through the bad luck of unsought isolation? Innately primitive or accidentally deficient? Perhaps we should ask a different question, one that neither Darwin nor Diamond considered: Why do any of us believe the Tasmanians were lacking in anything?

Maybe they couldn't make fire because they didn't need to. Jared Diamond's view that "warm clothing...would surely have been useful in Tasmanian winters" is a little like saying that central heating would have been useful in Tudor houses. The issue Diamond ducks is that doing something in a particular way entails something else. We tend to take such entailments for granted. A stove or a refrigerator entails a power source. It may come in a down-sized form, wind-generated or solar, but the core concept entails a nationwide power grid and more than a century of intense research and engineering that entails ongoing investment to keep everything functioning.

It is distinctly possible that the Tasmanians deliberately chose not to use bone fishhooks and chose not to catch fish, having chosen not to eat them (and so did not need the entailment of making axes to make boats to go to sea). This chain of choices led to a lifestyle that, while unexpected from a European maritime perspective, suited them. After all, they survived over 30,000 years of changing environment. We might be charitable enough to entertain the thought that the changes they made to their diet and material culture were refinements and improvements, making their way of life more, well...Tasmanian. On the surface, the degree of difference between the Tasmanian way of life and ours is so immense that even a nonracist commentator like Jared Diamond assumes that something must have gone badly wrong. If it was just the Tasmanians, we might believe Diamond. But Darwin's Fuegians (although they could make fire) were also stark naked in an equally cold, wet climate.

What if the Tasmanians decided not to make fire because they had decided to go naked? If not making fire and not wearing clothes were positive choices, then, plausibly, they were *connected* choices. On the face of it, dispensing with these key warmth-technologies is, as Diamond puts it, "folly." Before investigating what turns out to be the obvious connection, we need to understand how humans have adapted to different environments.

The archaeologist Robin Torrence, in a career spent comparing the living technologies of indigenous peoples with prehistoric tools, has noticed something highly significant about the range of objects people have, relative to the climate they inhabit. While peoples in tropical environments have—on average—more flamboyant, ornamental items than high-latitude peoples, she notes that their basic tool set is often more flexible and multipurpose. In contrast, Eskimo peoples, such as the Inuit and the Nunamuit, get incredibly tooled up for the onset of each season or phase of activity. Seal hunting requires a different set of tools from that for trapping arctic hares or for the salmon run. Because each activity is part of a regular annual cycle, people have several toolkits, each brought out at a particular time of year, used, cleaned up, repaired, and stored away until that season comes round again. In more tropical climates there is less seasonality and a greater abundance of resources, at least potentially. They are less predictable, and the challenge is to be prepared for many things. In a pinch, people in these environments can use the same bow to hunt and, strung more loosely, to make fire. Tropical dwellers are usually very good at "expedient technology." Expedient technology is what you reach for when you are stuck.

A proper angler would have laughed at me and my old digging mate Keith Moe, fishing in the Cascade Ranges of Oregon in spring. If we had been proper, fully equipped northern hunters, critically dependent on success rather than knowing that we were only a car ride away from Keith's freezer, we might have had a neat little metal-headed fish-killing club known as a "priest" (or, in Oregon parlance, a "fish whacker") with us. Instead, laying down my rod, and holding the slippery, wriggling fish in one hand, I reached around for an appropriate rock with the other.

Sacrificing smoothness for weight, and balancing a moral need for swift dispatch against my affection for my own fingers, I used three or four medium-weight blows. These unaesthetically but convincingly split the skull, knocking the eyes out. Things improved with the second and third trout, and the fourth was neatly sent to wherever trout go when they die (my stomach, I suppose). Afterward, Keith tossed our expedient artifact back into the water, and as the blood billowed off downstream, history evaporated. Unlike the rod, hook, and line, the improvised fish whacker reverted to being just another rock, unmodified and non-cultural.

With just two dozen manufactured artifacts, the Tasmanians lived in the flexible, opportunistic way characteristic of much warmer climes, although, since before the Ice Age, Tasmania had never been significantly warmer than it was at the time of European contact. Instead of looking after and transporting the gear they needed, they carried one or two tools and made anything they needed where they needed it.

There may, in fact, be a choice to be made. Although Robin Torrence is right to contrast the flexible responses of people in resource-rich, unpredictable environments with the highly logistical survival routines of those in high-latitude, harsh environments, the correlation is only general. As the archaeologist Everett Bassett has pointed out, the farther north or south you get, the more risk-reduction strategies are forced to diverge. The orthodox strategy is to become ever more specialized, going big on sleds, kayaks, harpoons, fall-traps, summer gear, winter gear, big-game gear, small trapping gear, and so on. As things become harder to find and hunt, water and wind get colder, and light and dark shift from a twenty-four-hour cycle to a twelve-month alternation. Investment in an insulating, adaptive technology is attractive. This is the "life-pod" approach, where getting food and staying warm are guaranteed by technological fixes at every point. The alternative strategy is a dramatic opposite and involves extreme opportunism. It is unorthodox, because in such demanding environments you need to be really good, divesting yourself of every encumbrance for maximum flexibility, weighing energy costs with potential risks at every moment. In the orthodox case it can be fatal if the *gear* fails, in the unorthodox case, if *you* do.

The dramatic and detailed black-and-white sketch of Hermann Buhl that I first saw reproduced in my boys' adventure annual exemplified the life-pod concept. Buhl was the first mountaineer to ascend a 26,000-foot peak solo (and remains the only mountaineer to have ever soloed a first ascent at this height). Having conquered Nanga Parbat as part of the 1953 German-led and -funded expedition, Buhl became an icon. Germany might have been defeated in the Second World War and carry the shame of the Holocaust, but, boy, could they make industrial stuff! Just as Austria had been annexed by its larger neighbor before the war, so Buhl, an Austrian, now became a figure of a greater German national pride. Schoolchildren were brought up on his image, in high-tech oxygen mask and cylinders, battling alone through a blizzard to reach the summit.

Many years after Buhl's death (in 1957 on Chogolisa), the slander was undone and the climber's true eminence revealed. He had made the ascent *without* oxygen in what Reinhold Messner (who made only the third ascent of the same hugely demanding peak in 1970, losing his brother) considers the hardest climb ever successfully returned from by a human being. Without heavy gas canisters for oxygen or for heating snow into drinking water, Buhl risked extreme exposure, altitude sickness, and dehydration. He pioneered the "alpine" style of Himalayan climbing, relying on immense endurance, skill, and speed instead of heavy-duty technological support. The German media saw the truth as shameful: Buhl, the Austrian maverick, had climbed the mountain like a primitive, and no one should know. And, ironically, the fact that Buhl did not carry a camera in his minimal gear made it much easier to create the myth of the conquest of nature by technology. Looking at Buhl's strategy—his personal alternative algebra of survival—we may doubt whether the Tasmanians "lost" much at all. Perhaps their nakedness created opportunities, not just for survival but for success?

Returning to the apparent mystery of the fish taboo, Robin Sim, who excavated some of the classic deposits with the fish bones, realized there was something odd about them.[24] Remember Diamond's conclusion that "Tasmanians used to catch many fish species, which accounted for about ten percent of their calorie intake"? A clue may lie in the "many species," as judged from the excavated assemblage of small bones—far more types than people usually target. More typical, in fact, of what spills out of a clubbed seal's stomach. Making a logical connection, Everett Bassett believes that the Tasmanians had no need for fishhooks or nets because they worked out that it was easier to let the seal take the risks, converting hard-to-catch fish protein into bigger slabs of mammal meat.[25] More work needs to be done to establish whether this is true, and some archaeologists think limited fish trapping took place, perhaps alongside seal hunting.[26] But no evidence for Aboriginal line fishing with hooks has turned up. And while the Tasmanians could have used bone-spear points for spearing fish, rather than further elaborating the bone-tool fabrication skills they had had 4,000 years ago, they really do seem to have abandoned bone-tool technology completely. Like many cultures who have changed the way they do things, they survived perfectly well. It obviously made sense to them.

The sudden disappearance of fish bones from ancient Tasmanian archaeological sites might better be interpreted as a change in seal butchery practice than an abandonment of the "let-the-seals-do-the-work" technique. The change may have been related to labor-saving and hygiene, as people decided to carry tools to slit open the gut at the kill site, rather than dragging the whole carcass back to base. Perhaps they took to eating seal on the spot, butchering it quickly and spit roasting the aromatic, blubbery steaks upwind of the stinking mess of semidigested gut contents.

The Aboriginal fish-eating taboo that so surprised the first European colonists now makes more sense. We all have food taboos. The English are unhappy eating horse, dog, guinea pig, fox, or insects and their larvae—all potentially nutritious (and available). The French eat the first, the Koreans the second, the Peruvians the third, Hungarian and Norwegian hunters (to my knowledge) the fourth, and a number of African communities the fifth. The reasons English people give for not eating these things vary. Some are pets, and some working animals—"honorary humans." Eating them would be almost cannibalistic. Others are considered dirty. Indeed, it is important to leach or "jug" fox meat and cook it very well to reduce bad flavors and protect against transferred parasitic disease. For the Tasmanians, perhaps the "I don't know where it's been" principle kicked in: as fish were what seals ate, and seals were what they ate, so fish could easily have been viewed with suspicion (Is it fresh? Or is it from a seal's stomach?).

Had Elizabethan sailors reached Tasmania in the later sixteenth century, we would have heard little or nothing of the indigenous fish foible. The loyal subjects of the Protestant Elizabeth I shunned fish too. The rejection was recent, symbolic, religious, and political—fish-eating was what Catholics did. In *King Lear*, Shakespeare's Earl of Kent brags of his utter reliability, confirming that he would "eat no fish." Elizabeth's key minister, William Cecil, thought the taboo was economic insanity on a large island rimmed by rich fisheries, and he promoted fasting days when a meat ban would, he hoped, force people to choose fish. In vain. It was nearly a century later, after the partial rehabilitation of Catholicism (and the brief restoration of a Catholic monarchy under James II), that a degree of honor returned to piscine cuisine. Meantime, English fishermen had been forced to diversify to stay afloat. Among other things, that meant piracy, buccaneering, "voyages of discovery," and, eventually, the colonization of places like Tasmania (by which time they were eating fish again).

What might be true of catching fish surely cannot be true of making fire. There was never a fire taboo in Tasmania, rather the reverse: fire was lit whenever a group stopped. The solution is right there, in what happens when something is so important that you organize it to the point where you don't have to think about it. I turn on the tap and it runs hot water. The Tasmanians were like that. They carried fire with them so that wherever they were, they could get a blaze going at once. The woodlands were full of flammable eucalyptus, and elsewhere there was plentiful pandanus, a plant that retains its old dry leaves beneath its fresh growth, providing a ready source of tinder. The benefit was speed and mobility; and they lit a fire whenever they stopped, however briefly. At times, this way of doing things became inconvenient. A band who lost their fire had to eat raw food and shiver until they found another group to beg a light from.

As a child in Britain in the early 1970s I remember "the winter of discontent," when the trade unions, led by the coal miners, took on the government. The working week was reduced to three days of available power supply. Some evenings we sat in overcoats and shivered, or fired up the Camping-Gaz to heat a can of beans in a saucepan. We lived deep in the country with a partly wooded garden. Did we go out there and make a splendid pit roast? No. It would have ruined the planting scheme. Suffering was what you did, and Brits were proud to be good at it. The excitement of deprivation passed, but power also returned. The village had a communal moan, society thus strengthened by adversity. Alternative solutions existed right in front of us, but that was not the point—we weren't anarchists, or hobos, and none of us was going to chop up the furniture, scorch the lawn, and have the neighbors think we had panicked. The psychological parallels might be close. Perhaps the Tasmanians *could* have made fire from scratch if they had absolutely needed to. After all, their inability is, as Beth Gott has recently argued, based on absence of evidence more than on evidence of absence.[27] Many hunter-gatherer groups carry fire logs about, as it is a much surer and faster way to have a campfire when one is needed. So the fact the Aboriginal Tasmanians carried fire does not prove they could not make it. Our understanding of their incapacity is based only on observations that indigenous tribespeople typically *did not* make fire. We do not definitively know that they *could* not. Maybe it was secret knowledge (and there may have been many things, in the circumstances, that they felt happier keeping to themselves). Gott's argument is on the side of attributing fully modern human intelligence to the Aborigines, but should not be accepted simply for that reason.[28] More pertinently, and more humanly, perhaps the Tasmanians simply considered having to make fire shameful, an affront to their established way of coping. Sitting stoically through the dark cold nights of the miners' strike, one could understand that.

There is no doubt that the Tasmanians, at the height of the last Ice Age, wore clothes. Could giving them up also have been a positive decision? In some parts of the island, especially to the west, the tribes built solid houses with insulated thatch roofs; on the milder and more sheltered southeast, the Bruny Island tribe, the Oyster Bay tribe, and the South-East tribe just had windbreaks.

What use are clothes with nowhere to store them? And how do you wash and dry them? Possibly, of course, you do not. In 1890, the Russian writer and medical doctor Anton Chekhov traveled to Sakhalin Island, between the Sea of Okhotsk and the Sea of Japan in the Russian Far East. As in Tasmania, convicts were being used as a colonizing force, displacing the native Gilyak (now known as the Nivkhi), who, Chekhov reports, "never wash, with the result that even ethnographers find it difficult to ascertain the colour of their skins. They never wash their underclothing, and their furs and boots look exactly as if they had just been stripped off a dead dog. The Gilyaks themselves exude a heavy, sharp odour...."[29]

We shall return to the Nivkhi later, and cast a critical light on Chekhov's understanding of their way of life, but there is no reason to doubt his factual accuracy. But, despite reaching some kind of endocrine homeostasis with their clothing, the Nivkhi changed their overclothes often, and they had roofed huts with hearths to dry them out. Wearing permanently damp clothes and boots saps energy; by speeding heat loss it can become a danger. Walking in the English Lake District, I have sometimes felt better to be barefoot, though I was bitterly cold, than to keep my soaked socks and boots on.

Had I thought to put some of my dubbin (old-style leather shoe grease) directly on my feet, I might have been more comfortable. Body grease is something long-distance sea swimmers use in preference to a wet suit. It's an issue of acclimatization, and a choice of discomforts. If you can adapt psychologically and physiologically to a life without clothes in a high latitude, then you are freed from an awful lot of sitting around, weaving, sewing, mending, and, of course, getting damp and cold.

Perhaps this explains the expedient technology of the Tasmanians. Instead of sitting down for a long time to make a complex tool that you might lose or damage, you hardly break stride to knap a sandstone blade edge and deal with that seal. The Tasmanians were highly skilled land hunters, yet they used neither spear thrower nor stone-tipped projectiles. They did not have ground stone tools because grinding stone is very laborious, whereas efficient knapping can be a matter of a few highly skilled strikes. Everything was quick, and replicable. If a blade was lost, you made another one, or picked up an old one and refreshed the edge. Being without clothes reduced your other possessions, so that what you owned was yourself. Entailment was minimized. This was Hermann Buhl's logic on Nanga Parbat: not naked, but with an absolute minimum of gear. It could be described as *reverse entailment*.

Most modern technology is very highly entailed. A car needs wheels and fuel. Those entail rubber plantations and oil wells, and complex manufacturing, refining, marketing, and distribution processes. Once all the things that cars have to have to be cars are factored in—from metal, tarmac, and glass, through to traffic police, a licensing bureaucracy, test agencies, and so on, each of which comes with its own primary and subsidiary systems of entailment—it is clear that the car can exist *only* within a modern globalized industrial system. Reverse entailment, although it sounds as if it might be even more complicated, means a type of dissociation, a deliberate unmeshing. To our habitually entailed existence, the prospect seems strange, the opposite of progress.

Here is the theory of reverse entailment, presented as what is *not* entailed by the Tasmanian Aboriginal way of life: no bone tools means no awls means no clothes means no pockets means nowhere to keep tinder and fire-making kit; that means no fire making, which means carrying fire all the time. That means quick fires whenever you want, which means it is okay to have no clothes. Carrying fire all the time means it is safer not to wear clothes, so you don't catch fire by accident. It also reduces what else can be carried. No composite tools means no axes, so no log boats or all-weather craft, which means no fish—but with lots of sudden storms, why risk orphaning the children when seals can fish and you can eat seals? Inshore, you can grease up and dive for lovely stuff. Naked, of course, because you don't want damp clothes. You want to get dry and warm as fast as possible by the fire, and eat abalone, wallaby, and tree-harvested possum. Your little temporary rafts can take you to small seabird breeding haunts in fine weather, and who wants to go birding and egg-hunting in high seas? Without a complex toolkit to lose, or surpluses to be stolen, or clothes to dry, you don't really need a house, and since you carry fire everywhere, the risks of burning one down would be high. With no houses and little personal property, there's not much hierarchy. That means no need to make loads of stuff, like thrones and crowns. Accounting is unnecessary, so you don't need writing or numbers, or pens or paper. Or money. You have no maps, but you are not lost. You know where absolutely everything you need is. And because you don't have to look after it, you can get it when you want it.

The more we look at the Tasmanian Aboriginal toolkit, the less the parallel with the tools of chimpanzees (legitimately enough made on formal grounds by Bill McGrew) makes sense. It is not just that the humans had more things, because with only two dozen items, it was not that many more. It is that their technology was not an add-on, an optional extra. It was essential and embedded. Chimps can live without tools. Humans cannot.

The Tasmanians used a set of carefully chosen, long-tested essentials, honed down from a repertoire that was once far broader so as to retain (as in Buhl's Nanga Parbat ascent gear) absolutely nothing unnecessary. Well, almost. Because, as well as the fire sticks, wallaby-hide infant carrying slings, kelp-frond water buckets, wooden spears, clubs, digging sticks, twisted fiber ropes and baskets, and animal-skin pouches, they made perforated shell necklaces. Like Buhl's Fair Isle–style wool sweater, their body painting and hair treatments showed a certain somber aesthetic.

By dramatically paring down their basic equipment, the Tasmanians lost some abilities that outwardly similar high-latitude hunter-gatherers retained. But other human groups in cold environments may have existed like this too. One prehistoric example seems to be the Dorset culture, which preceded the Inuit in the American North. The Dorset had a much simpler technology than those who came before or after, "losing" their bow and arrow technology, and their abilities to drill bone and make complex, composite harpoons and sophisticated watercraft. On the face of it, they were a "degenerate" culture, and that is how (without using that precise pejorative) archaeologists have usually viewed them. Yet the Inuit, who drove the Dorset to extinction, have legends telling how hard it was to defeat these consummate hunters and brilliant survival strategists. The Yahgan people Darwin encountered in Tierra del Fuego could easily be seen to fall into this "degenerate" camp too. Like the Tasmanians, they used marine mammal grease for insulation, and slept naked without shelter, and among them only the women swam. This feature was probably more than an arbitrary division of labor. In such extreme cold waters, a female pattern of body insulation may have been essential. Jared Diamond notes this very same pattern among the Tasmanians; he believes the Tasmanian women were operating at the absolute boundary of human physiology, beyond what men, with their thinner subcutaneous fat, could survive.

The Tasmanians had evolved their skills for over 30,000 years, cut off from the continent for the last 10,000. That final episode gave birth to a minimalist survival style that was highly refined, in the sense of being purged of anything unnecessary. What Darwin and others saw as innate incapacities, and Diamond recasts as unfortunate losses seem to have more realistically been considered distractions from the skills they really needed. Women who had learned to dive naked, in waters that some hi-tech, wet-suited, diver might balk at were women who wanted a fire and wanted it *now*. They did not want to emerge from the depths to find someone messing around with sticks and stones and damp tinder.

The collective Tasmanian achievement was, like Buhl's solo ascent of Nanga Parbat, almost superhuman. At the time of European contact, the Tasmanians had reached a point where they not only could do their stuff without clothes, but could do it *better* without clothes. Ironically, for me, not making fire is a badge of modernity; for them it became a mark of abject savagery, one that persists.

Although they seemed to live very close to nature, they required critical and constant insulation from it, based on the core techno-cultural patrimony of our species. They killed with spears, cooked food with fire, and despite their corporeal nakedness, applied specific protections and cosmetics and transported their children in slings. Their stone tools were expedient only because it made sense for them to be. Unlike chimpanzees, who can and do live wholly in nature, the Aboriginal Tasmanians were as critically dependent on technology as were more mainstream artificial apes.

As native Tasmanians became rarer, interest in them increased. Genuine skulls fetched between five and ten shillings in the early 1800s, and there was fierce competition among learned institutions to obtain good examples of this "evolutionarily primitive" grade of humanity. Only in 2002 did the Royal College of Surgeons of England relinquish skin and hair taken from "Queen" Truganini—the last woman in the world from a culture too ignorant to make fire. Made sterile by syphilis, the childless Truganini feared as she lay dying that she would be dissected as a zoological specimen. She asked to be buried in the bay where her mother—murdered by whalers—had given birth to her; there, or alternatively, in her last recorded words, "behind the mountains."[30] Instead, her bones were boiled clean, wired back together by the Tasmanian Museum and Art Gallery, and placed on display not far from a kangaroo skeleton.

Chapter 3

UNINTELLIGENT DESIGN

With a small axe, I hacked off a couple of haunches, leaving the rest to rot. These creatures were simply crude, unrefined machines, with a limited lifespan. They had neither the robustness nor the elegance and perfect functionality of a twin-lens Rolleiflex, I thought, as I looked at their protruding, lifeless eyes.

—Michel Houellebecq, *La possibilité d'une île*[1]

I CANNOT TELL WHAT he is thinking as he stands motionless, his pupils shrinking into pre-attack blankness. His nostrils dilate at the moment he swings for me, the massive force and acceleration raising his whole body into the air. There is no time to flinch, but the single punch that might have knocked a Joe Louis or a Mike Tyson out cold rebounds off the inside of the security screen. He retreats as suddenly as he came, collapses against the far wall, and, with a studied nonchalance, begins to pick out straw from his fur.

Darwin felt kinship for creatures like this, considering them kin. Not socially close, certainly, but properly related. For many people brought up on the story of Adam and Eve, experiencing the world as flood-washed and gleaming, a mere six millennia old, Darwin's claim was—still is—a scandal. Even those who saw the dramatic changes that selective animal breeding brought about among dogs, cattle, sheep, and horses balked at accepting apes into the family. Doing so would, ultimately, logically, mean having to admit said farm animals into the fold too. Of course, Darwin's contention was just that, that all life on the planet was genealogically related if the relationships were tracked back far enough, humans to hardwoods, fungi to fish, ants to apes.

As we now know, through correlating the fossil data with estimated rates of change in mitochondrial DNA, to make any sense of the claim of our close kinship with apes, we have to presuppose not 6,000 but at least 6 *million* years of subsequent divergence.[2] And we can now estimate the approximate number of great-great-greats in the appropriate grandparent term. It is a quarter of a million. That is the minimum number of generations between me and the last ancestor I held in common with the alpha male gorilla. But a different kind of separation concerned me in Birmingham's Twycross Zoo: a practical one, dependent on a few crucial millimeters of superbly engineered glass.

Darwin's initial impetus for becoming a natural historian was reading the *Natural Theology* of William Paley, who thought that the mechanisms of the natural world displayed complexity exceeding that of the finest watch "in a degree which exceeds all computation."[3] God, for Paley, was the Great Designer, proof of whose genius was manifest in wonders of engineering such as the spine of a hare. Paley wrote, "the spine, or backbone, is a chain of joints of very wonderful construction. Various, difficult and almost inconsistent offices were to be executed by the same instrument. It was to be firm, yet flexible... firm, to support the erect position of the body; flexible, to allow of the bending of the trunk in all degrees of curvature."[4] Paley believed God had achieved this through the use of stabilizing fins—vertebral condyles—which held the disks in line and protected the spinal cord, while allowing muscular flexing in almost any direction. These condyles Paley thought he recognized as "the very contrivance which is employed in the famous iron-bridge at my door at Bishop-Wearmouth."[5]

Darwin, too, was fascinated by the mechanics of anatomy, yet he came to believe that the only "design" had been produced by trial and error. Even he, though, was inclined to downplay the degree of dysfunctionality that evolutionary compromises might produce. One of the most unintentionally misleading images in all evolutionary science is of the knuckle-walking ape slowly becoming an upright human. In fact, the transition from knuckle-walking or tree-swinging to bipedalism requires an abrupt shift in posture. Whether it is an individual attempt, or a postulated evolutionary change, no creature can hang around long in the stooped phase without getting a very stiff back. But there is something more misleading in the way in which Darwin and his followers demonstrated the comparison between the great ape skeleton and the human one. An example is on the front dust jacket of James A. Secord's celebratory 2008 collection of Darwin's *Evolutionary Writings*.[6] There we see two skeletons, one a gorilla (to the left), apparently trying to stand up, and (to the right) *Homo sapiens* walking away. Of the two, the human looks far more graceful. That is not simply a species bias in my aesthetic; rather the gorilla is placed in an uncomfortable position. But, if we look at this great ape's bones articulated naturally, with the rib cage slung beneath an almost horizontal backbone, the fine-tuning of natural selection becomes clear, and it is now the human that seems strange. As indeed it is: the spinal column, perfectly suited to suspending a rib cage, has become a teetering pillar continually pulled forward and off-balance by the thorax, while supporting a ludicrously large cranium.

The inadequacies of human backs are just the beginning. If a designer had ever been involved, then, scrutinizing their work, we could reel off a litany of one utterly mad design decision after another—inefficient digestion, bad insulation, tender feet, fragile nails, feeble teeth, poor sense of smell, and weak vision.

Ancestral apes used things they found, like stones and branches, as tools and weapons. But, although they could crack open nuts without breaking their teeth (mainly) and hit their enemies and each other quite hard without hurting their own knuckles, they were not intelligent enough to make the next move—to transform the natural objects they found. They lacked the lethal ability to chip a stone into a point, attach it to the end of a smoothed branch, and lie in wait. It is usually thought that human beings slowly evolved from apes and, at a particular point, our brains became so large that we were able to make our own tools, each fit for its specific purpose. In this scheme, biological evolution had given us upright walking and much bigger heads than our rivals; now, cultural evolution began, leading ineluctably from flint axes to silicon chips. We did not need to evolve physically anymore: the era of technological civilizations had begun. But none of this makes as much sense as the opposite scenario. That is, if instead of our becoming intelligent enough to invent things, the things actually allowed us to evolve into intelligent human beings.

There is no accurate measure of gorilla strength. Chimpanzees, as estimated by the Jane Goodall Institute, are "at least five times" stronger than most humans, while orangutans, as recorded by Alfred Russell Wallace in 1856, are known to rip apart the jaws of saltwater crocodiles, killing them with their bare hands.[7] The male gorilla is around three times heavier than a chimp and twice as heavy as a male orangutan. It can be reckoned at between eight and fifteen times stronger than an average adult human. Thus when humans and gorillas meet by mischance, the dice are heavily loaded. In 2004, a silverback called Jabiri jumped a fourteen-foot-high wall at the Dallas Zoo and went on the rampage, trying to eat the head of a three-year-old toddler before throwing him at a wall and turning his attention to the boy's mother. Both humans survived after hospital treatment for near-critical injuries. In 1998, a gorilla called Hercules attacked an intern keeper at the same zoo, biting her arms and legs. And in Chicago in 2005, Lincoln Park Zoo's gorilla, Kwan, carried through a dominance display on a female zookeeper by knocking her down and biting her back.[8]

Perhaps to counter the King Kong idea (the original 1933 Merian C. Cooper brute, not the sensitized-to-the-ambiguity-of-modern-masculinities Peter Jackson remake), primatologists are keen to stress that gorilla nature is "essentially peaceful," at least away from the psychological pressures of captivity. But silverback males establish peace on their own intimidating terms. Harems are built not just on the threat of violence—chest beating and the flashing of massive canines—but on a willingness to carry through on displays. In the wild, males frequently knock down and bite females they wish to possess and have been observed to rip baby gorillas from their mother's arms and hurl away the offspring of rivals in order to assert their power.

They may also attack such rivals directly; a rare event, but the results are not pretty: Diane Fossey, the highland gorilla researcher who became the subject of the biopic film *Gorillas in the Mist,* once found a gorilla skull with a canine punctured through the top. Lynn Kilgore of Colorado State University has studied gorilla and chimp skeletons extensively for signs of violence and finds that 22 percent of gorillas and 23 percent of chimpanzees show at least one skeletal trauma. Much of the damage consists of fractured cheekbones and missing teeth, consistent with aggression.[9]

I am not detailing zoo attacks and drawing attention to the periodic use of violence in the wild so as to re-demonize gorillas or chimpanzees.

Shot for bushmeat, their habitats raided for firewood, their skulls and skins sold as trophies, these amazing creatures are on the verge of extinction. However, to fully understand why that is, we need to understand what has made us so dominant, such a potential threat to so much planetary life. Put another way, how is it that the most naturally weak of the great apes, the human species, has emerged, after all, as the most powerful? How did we reach the point where Jabiri's Dallas Zoo rampage was finally cut short by a burst of lethal gunfire from police marksmen? Could it be that the very strength of the male silverback stands in the way of an ability to manipulate and invent? But, if so, how did our ancestors manage to trade natural for artificial strength? Did we stop being ape in gradual stages, or was there a critical moment? To attempt an answer we will have to try to understand what our diminutive evolutionary ancestors were up against in their original struggle for survival in the wild.

Staying alive was a demanding practical project for our tiny ancestors. They succeeded, but a whole series of questions arises about how. Somehow they had to find a way to emerge from the shadow of such powerful creatures as gorillas, lions, leopards, saber-toothed cats, and large birds of prey. We know from the remains found in saber-toothed *Dinofelis* dens that *Australopithecus afarensis* were on their menu.[10] And we shall shortly turn to the deeper implications of what befell an australopithecine child carried off by an eagle.

To understand what is going on, we first have to take a close look at what our innate biological capacities actually are. We need to take a look at the shape we are in and understand how very different it is from the shape that we could have taken. That story begins wholly in nature, with the dynamics of speciation from a common ancestor, which is just another way of comprehending how one species can become two subsequent species, different from their ancestor and from each other.

I recently took two fish, both dead, out of my briefcase and passed them around an audience of schoolchildren. One was a mackerel—silver, streamlined, fast. The other was a brill, a member of the bottom-dwelling flatfish or flounder family. The fish were fresh from the fishmonger's slab, ungutted, wet, cold, and fishy smelling, and they changed hands to gasps, giggles, and some squeals. As they went around, I drew attention to their eyes and gills. The mackerel's were to either side of the head, perfectly symmetrical, just where you might expect them to be. By contrast, on the brill both eyes were on the dark upper surface, poking up in a strangely twisted manner. There was a single, functional gill behind them, and I encouraged the children to look underneath and spot where the other gill was. Some saw it, on the white underskin, looking like a small, crescent-shaped wound, atrophied and hardly functional. Holding the brill upright and looking into its mouth, the children could all see the warped resemblance to a free-swimming mackerel.

Baby brill start out just like mackerel. Their eyes—at least the cell bundles that will become eyes—start out one on each side. The gills are equal in size, the head is symmetrical . . . it looks like a regular, "normal" fish. But, at a certain point in the fetal development, one eye migrates right through the head and pops out at a funny angle at the side of the skull, next to the other eye. This feature gives the whole flounder family, from plaice to turbot, halibut to lemon sole, the ability to lie flat on the seabed, effectively on their sides, catching shrimp without loss of binocular vision. It also gives them a rather ridiculous drunken frog squint.

I wanted my fish demonstration to make the point that evolution is a fact that is observable all around us, not just a powerful theory. The child who asked me, "If we evolved from chimpanzees, then why are there still chimpanzees?" set me thinking that the way the evolutionary story is presented is often confusing. The brill is *not* the descendant of the mackerel, in particular, but it has diverged from a round fish that looked much more like a mackerel than a brill. The brill modified the original plan dramatically. Similarly, humans descended from apes, but so did apes. Modern gorillas are descended from an ape ancestor that was not a gorilla and that was also our ancestor. But that ancestor was more like a gorilla than we are like a gorilla. Our subsequent modifications have been more dramatic, and are hardly more elegant than those of the brill. We may not have ended up with drunken frog eyes, but the traces of radical compromise, organs evolved to one purpose and then more or less crudely adapted to another, along with a story of loss of biological function, are written throughout our frail bodies.

As Darwin looked at his careful tally of scores from the regular backgammon games he played, often twice a day, with Emma, a pattern emerged. The board allowed for both chance and skill, and neither was a bad player. But, over the years, Charles recorded slightly more wins for himself. He did not attribute this to an individual quirk. Nor did he seem aware of the likelihood that, through a series of arduous pregnancies and the running of a large household, with its budgets, internal diplomacy, periodic crises, and employment issues, Emma may sometimes have assented to her husband's desire to play at times when her mind was on other things. Her diplomacy might even have extended to letting him win from time to time, but we will never know. What we do know is that Charles Darwin assumed that the average of backgammon games in his favor indicated that Emma's brain was slightly inferior to his in just the way that his theory of evolution and, particularly, his understanding of sexual selection in mammals would predict. His prejudice was that men found physical beauty attractive in women but were not themselves objectively desirable; what a woman wanted in a man was brains, so that she could benefit from his clever solutions, and pass his ability for advanced know-how down to their offspring.

But although Darwin thought women were naturally more attracted to intelligent mates, this was not what he was attempting to explain. In theory, women should have been able to choose intelligent mates while remaining deeply unintelligent themselves. The peacock's tail example showed that among peafowl, the hens had no pretty fantail but chose those male mates with the most elaborate ones. Fantails were sexually selected, and the finest tails passed down from fathers to male offspring. If human intelligence, as Darwin thought, had also been sexually selected, then why was there no such dramatic contrast between the mental attributes of men and women? The peacock's tail explanation of the human brain left Darwin with a puzzle. Accepting that his own highly elaborated brain was a sort of evolutionary sex signal, Darwin wondered if his wife, Emma's, brain should not, with equal logic, be equivalent to the tail of the peahen, small and comparatively undeveloped? Clearly, no such immense discrepancy existed... but then, neither was Mrs. Darwin necessarily *quite* as smart as her husband.[11] What Darwin wanted to understand was why, among humans, the intelligence of a woman was—in his own firm opinion—almost, but not quite, the equal that of a man. Let us assume, for a short while, that this opinion is justified, and see where Darwin's logic leads.

Darwin had noticed that in most insects and many birds, sexual selection had produced major differences between the sexes. Some male and female members of the same species were so different in appearance—color, shape, body mass, and so on—that an earlier generation of naturalists had assigned a fair number to entirely different species. But in mammals the effect was far less marked. Males were often a bit bigger than females, and might have larger horns or fangs, and perhaps a bulkier ruff of fur. But in many species it was quite hard to tell males and females apart. Today we know this is because genes that have been selected as conferring an advantage for one sex still find expression in the other sex, in a process known as pleiotropy.

In general terms, pleiotropy has to do with mammals being later products of evolution, their complex and sophisticated bodily features being coded for by broad sets of genes. Because very few features are coded on just a single gene, sex-specific peculiarities get diluted. There are some specific and focused sex-specific traits in mammals, as well as genetic faults that affect one sex more than another (such as color blindness in humans).

Darwin did not know about genes and described pleiotropy as "the law of equal transmission of characters."[12] It was with some relief (but perhaps a little male self-satisfaction too) that he wrote: "It is, indeed, fortunate that the law of equal transmission of characters to both sexes has commonly prevailed through out the whole class of mammals; otherwise it is probable that man would have become as superior in mental endowment to woman, as the peacock is in ornamental plumage to the peahen."[13]

Examining a gorilla's jaws firsthand in the company of a world authority on biting teeth, Dr. Peter Kertesz, dental consultant to the London Zoo, was an education. Kertesz's consulting room houses a fascinating collection of animal skulls and specimen animal teeth. The outwardly bizarre reason I was there was because I was involved in a Discovery Channel documentary on the science underlying different aspects of the classic Hollywood vampire myth.[14] Having already investigated subjects ranging from hematology and eastern European folk beliefs to the grizzly crimes of obsessed serial killer psychopaths and eccentric practices on the New York fetish scene, a key question remained unanswered, one that I hoped Kertesz could help with.

A case we had covered earlier in the filming concerned the "Vampire Rapist of Montreal," Wayne Clifford Boden. In 1971, Boden had the dishonor of becoming the first person in North America to be convicted using forensic odontology.[15] The idea of using the peculiarities of teeth as a personal identifier has a long history. The emperor Nero kept clay impressions of bite patterns in order to keep track of his palace slaves. In Boden's case, bite marks on the breasts of one of his victims were matched to distinguishing features of his own dentition. But what I needed to know was whether Boden's bites were anything like the ones in the Dracula films.

Early on in Bram Stoker's gothic chiller, *Dracula*, Mina Murray uses a "big safety pin" to secure a warm shawl at Lucy Westenra's throat. The friends have just experienced a nighttime scare in East Cliff cemetery; unbeknown to them (though not the tremulous reader), Count Dracula has made landfall and begun to prey on young women. When Mina wakes Lucy in the morning, she notices something that makes her blame herself for lack of care. Stoker has her write in her journal: "I must have pinched up a piece of loose skin and have transfixed it, for there are two little red points like pin-pricks, and on the band of her nightdress was a drop of blood."[16]

Thus was born the cinematic image of the twin wounds on the vampire victim's neck. The answer from Kertesz as to its plausibility was, unsurprisingly, negative. Even were a psychopath, playing the evil count, to wear prosthetic teeth, such a pattern could not result. As we spoke, sitting in his dental chair, dressed in full Goth gear, was a vampire fan called Tara Danes, who was considering having permanent "vampire" fangs implanted. She already had two pairs of custom-made—and surprisingly sharp—snap-on fangs. One pair lengthened her upper canines for the typical "Dracula" look; the other fitted more centrally over her upper lateral incisors, on either side of the two front teeth, to produce the "Nosferatu" look. Using the skulls of a lion and a gorilla, Kertesz demonstrated the obvious: that to get the classic paired Hollywood puncture marks there would need to be corresponding marks from lower canines. The mechanics of leverage demand it. Kertesz then graphically explained to Tara why he, at least, would not be prepared to grant her wish, although less scrupulous dentists are known to carry out the procedure.

Looking at a gorilla skull juxtaposed with that of a real person, the architectural differences leap out. The creature that seemed so like me when it stood upright to stare me in the eye, six feet tall behind the glass, has a skull that is in many ways more like that of a dog than a human. The gorilla has a muzzle, or snout. This projecting aspect, technically termed prognathism, is necessary to house the upper canines, which are much longer than one might imagine. Even having examined a removed canine, it is still hard to accept the evidence, namely that what is visible in a living primate, whether human or ape, is only the lower *third* of the tooth. Precisely because this is a ripping tooth, it has to be very securely anchored. Hence the gorilla's muzzle is actually mostly an anchor platform for the canine roots that rise through the bone on either side of its flat nose. The actual root tips, buried in bone, are not that far from the inner corners of the eyes.

Unlike humans, the other great apes also have a diastema, or gap, between the incisors and canines at the top and the canines and molars at the bottom. This allows the upper and lower canines to slide past one another and snugly seat themselves so that the jaw can shut properly. Had Kertesz agreed to fit Tara with a permanent prosthetic enhancement to her own upper canines, he would have been endangering her health on two counts. First, the newly lengthened canine would not be anchored by a long enough root. Second, the lack of a lower diastema would mean that she would be in danger of being unable to close her jaw fully, and also perhaps puncture her own lower lip—about the only way a Dracula cultist could be left with Stoker's emblematic "two little red points."

For a gorilla to bite effectively with its big canines, the lower jaw has to close hard against the upper. This is achieved by muscles, especially the paired temporalis muscles, which pull the open mouth shut. The jaws need to be pulled shut with great force if the mega-canines are to present any real threat. That means that a big muscle needs to be attached under the gorilla's lower jaw, the other end of which needs to be attached to the upper jaw to pull the two sides together, closing the hinge. But this will not work because the upper jaw is in fact the muzzle—a tube that presents a smooth curving surface onto which the top end of the jaw-closing muscle cannot properly attach. Worse, this surface is punctured by the delicate nasal opening right in the center. If the gorilla's muscles were somehow attached here, then the animal would be in danger of ripping its nose apart whenever it tried to bite. In reality, the temporalis muscles rise from either side of the lower jaw and run diagonally backward and upward to form part of the cheeks. But they are not attached there; they pass on up, to either side of the eyes, and firmly connect to the very top of the head where there is a dramatic raised crest—a sort of bone fin—that runs over the top of the skull, front to back. Power is thus triangulated from the point of origin of this sagittal crest, just above the eyes, through a mesh of bone and muscle, and terminating in the deepest anchor for the canines, just below the eyes.

The sagittal crest is found in many predators, including the tyrannosaurs, and extends even further back in evolutionary time to the pre-mammalian cynodonts, the first creatures with canine teeth, which lived some 250 million years ago, in the Triassic period. The presence of the crest and temporalis muscles accounts for the powerful bite of modern canids, such as the rottweiler dog, the gray wolf, and the pit bull terrier, which generate jaw-closing forces of 800, 1,500, and 2,000 pounds-per-square-inch (psi) respectively.[17] The gorilla's bite works the same way, and although its bite force has not (to my knowledge) ever been precisely measured, from muscle bulk and the size of the sagittal crest, it can be thought of as falling somewhere within this range. Ignoring his fictional status for a moment, the vampire Count Dracula would not stand a chance against any of these creatures in a bite contest. Not only would his canines, without a properly deep root in his flat, muzzleless (non-prognathic) face, break off, but his bite force, in the absence of a sagittal crest, must essentially be the same as ours: a mere 200 psi. That works out at somewhere between a quarter and a tenth of the gorilla's.

As Tara swapped her canine fangs for her laterals, I felt a curious change in my own responses. While the laterals looked distinctly spooky, the canines were actually scary. It was a few months later, after a visit to the Brazilian rainforest, that I finally settled on why this may have been so. The Amazon is the home of the vampire bat, so named by white settlers because it evoked Central European legends of nocturnal blood drinkers. No such bats exist in the Old World. The naturalists I was with were monitoring the spread of rabies, and after nearly being bitten by a captive bat (a very fast mover), I was able to get a close look at its teeth. Its little pair of razor-edged upper teeth are not canines and do not puncture (so they cannot produce the legendary twin puncture marks either). Its fangs slice the surface of the host's skin back with surgical (and apparently nearly painless) precision, so that its anticoagulant saliva (active ingredient "draculin," what else!) can dribble onto a fresh wound, allowing fluid blood to be licked up and swallowed. This leaves the bat so full that it cannot get airborne to return to its lair and has to waddle away over the forest floor.

Tara's second set of teeth mimicked those of the vampire bat, but at one remove. Although vampire bats do not exist in Europe, Bram Stoker implied that his evil Count Dracula could adopt bat form. The vampire idea, carried to Brazil to label the blood-drinking bat, had reverberated back across the Atlantic to graft itself back onto the parent vampire myth, enhancing the already present shape-shifting "creatures of the night" aspects of the legend. When the German expressionist F. W. Murnau made his silent film *Nosferatu*—a non-copyright version of Stoker's story—in 1922, he gave his "Count Orlok" the very long sharp central teeth of a vampire bat. Orlok, played by the mysterious mime artist Max Schreck, certainly has a deeply chilling aspect. But he remains a nightmarish and otherworldly horror. When Universal Pictures shot their official *Dracula* in 1931, Bela Lugosi was given more widely separated fangs in the form of prosthetic upper canines.[18] These produce a believable sense of real-world threat.

It is a fact of human evolution that, in the course of it, we lost both our sagittal crest and our large canine teeth. That is why we cannot bite very hard in comparison to a gorilla. It may also be why vampire fans get a kick out of wearing longer canines. The bared canine display is an evolutionary throwback, a link to our bestial past that stirs some deeper regions of our brain—not so much an "ancestral memory" as a hard-wired response to a snarling opponent. As Tara swapped her Nosferatu canines for her Dracula ones, her appearance shifted from whacky and unsettling to first-cousin primate aggressiveness, at once more familiar and more intimidating.

So why did we humans lose our canines? The answer appears to be simple, but turns out to be anything but. Making the *Real Vampires* program was an opportunity for me to revisit Transylvania, that part of modern-day Romania where Stoker set his story, and on whose mountain border the real Vlad Dracula—not a vampire but a notorious fifteenth-century warlord who impaled his enemies—had once lived. The trip in the summer of 2005 brought back memories of several months in 1983 when, as a young postgraduate student in archaeology, I had been part of a project studying the traditional life-ways of upland pastoralists. The conditions had been as tough as the heavily forested mountains were beautiful. My tiny expedition tent attracted a wolf one evening, and often a pair of us would find ourselves deep in boar or bear territory. The Transylvanian proverb goes "if you go hunting bear, take a doctor; if you go hunting boar, take a priest," but in practice the major threat was from humans and their dogs. Each summer hut (a seasonally used shieling, or *stîne*), set in the forested flanks of the Carpathian Mountains, was occupied by a different family. Each family had a selection of mastiff dogs, mostly of medium-to-large size and many with spiked iron anti-wolf collars. It was alarming to see the crofters throw stones at their own dogs to intimidate them and keep them from mauling the children. It was more than alarming when one day a fellow student and I stumbled over a croft not on our local sketch plan, belonging to the local family from hell, who were feuding with all the others and had consequently not been included in the list our guide had provided. We were welcomed by their massive, spike-collared mastiff dogs, barking in quadraphonic, and slavering so much that I had an interesting opportunity to study their canine teeth.

When we consider our human ancestors, out at night on the African savannah, we have to be practical. Personally, my blood runs cold when I try to imagine helpless human ancestors not just without security glass, but without guns or bows, torches or fire, knowing that there were not only much bigger apes wandering in the dark, but lions, leopards, hyenas, rhinoceroses, crocodiles, and elephants. Frankly, if I had young children with me, I would be concerned about spending a night out on the African savannah even with a campfire and a gun.

That our evolutionary ancestors started out far from the top of the food chain is graphically illustrated by the Taung child fossil, the remains of a little *Australopithecus africanus* who, some 2.6 million years ago, was eviscerated and carried off by a crowned eagle.[19] For some reason, the eagle did not finish its meal; although the brain had been exposed, the remains fell in among debris beneath the eyrie, and an endocranial cast was miraculously preserved, punctured by distinctive talon marks.

So it is in practical terms that we have to try to grasp how our ancestors were able to emerge from apedom toward civilization, discarding the massive canines and losing muscle-mass in such a threatening environment. It is all very well to say that they managed it by inventing chipped stone axes and all the rest of it, but those are things an ape is not capable of inventing without *first* losing many of its natural defenses.

This is because to get a more intelligent brain in a human ancestor, the cranium has to become both larger and less robust. Instead of the heavy doglike head of the gorilla, we have a skull in which the snout has gone, the face being vertical, tucked in under the eyes. Prognathism goes and with it the large canines. These would no longer serve any purpose anyway, as the high-domed skull can no longer support a sagittal crest, so the jaw muscles become a lot weaker.

Holding a prehistoric baby's skull, it is impossible not to look into its eye sockets. They are not as dark as the sockets of adult skulls; the cranium was stretched so thin in its headlong growth that the vault is translucent. Once, light passed in only between blinks, swelling the bubble with names, schemes, affections, and hopes, all the things that lent a dissembling tautness to the now empty forehead. But look into the eye sockets of the skull of a newborn gorilla and the effect is comparable. Indeed, if you did not know, you might think that the ape infant was human, so similar is its initial appearance. But, whereas a human skull retains many childlike features while expanding hugely after birth (processes of neoteny and paedomorphism that we shall shortly turn to), the gorilla's skull rapidly becomes robust. The human cranium remains delicate, high-vaulted, and thin. Some geneticists believe that an alteration in the gene MYH16 weakened the development of the jaw muscle and actually allowed physical pressure to be released from the top of the ancestral human cranium, causing the rapid expansion of brain size and reasoning power that we can track in the fossil evidence.[20] Whether that turns out to be correct (even in part), the fact is that I cannot easily do what a chimpanzee does: eat uncooked meat straight from the carcass of a freshly killed monkey (or dead chimpanzee infant from the same troop). Relatively speaking, I have the smallest canines of any primate. I prefer my meat cooked, and cut up, both before and after cooking. And if I kill, I use a tool—a club, gun, or bow of some kind—to amplify and focus my strength and insulate me from the physical dangers of ripping up something at close quarters (perhaps even to insulate me psychologically . . . but that argument must wait).

The standard answer to the question of how human evolution took place is that our direct human ancestors, starting out as tree-living niche specialists scavenging for scraps on the margins of the main action, gradually got smarter. Intelligence, we reason, reasonably and intelligently, somehow benefited us in our social organization. It made us more similar to what we were to become: good at communication, forward-planning, and finding safety in numbers. Thus, our upright-walking hominin ancestors were able to outwit not just other apes, but every other predatory carnivore and belligerent herbivore on the African savannah. Breaking out of that great continent as the big-brained genus *Homo*, we did battle with mammoths and cave bears, marsupial lions, and saber-toothed cats. Reinventing ourselves along the way, by degrees we became the top predator in almost every ecosystem, colonizing Eurasia, Australasia, and the Americas, helping to drive many large mammal species to extinction.[21]

But however it is that genes interact with mechanics either to cause or to allow the architectural changes that certainly took place in the evolving human skull, it is clear that we now cannot live as other apes do. We actually have to have complex tools in order to get away with our—in nature's terms—pathetically underdeveloped jaws. Armed with these tools, we can feed on the very high levels of protein that our big brains need for proper functioning. But how could we have invented tools if we started out stupid? And how could we have been successful enough in their absence to be able to evolve ever larger brains?

The central paradox of our existence is that we are the product of the artifice that we ourselves brought into the world. Un-invent the security glass and all the rest of the technology that augments my relatively feeble frame, and it is hardly conceivable that I would be able to compete with gorillas and chimpanzees in the wild. Yet so far has my species attained the upper hand that we capture great apes alive and fly them in pressurized containers across oceans and continents to populate purpose-built environments where they can entertain, educate, and, increasingly, be briefly protected from yet another human-driven extinction.

Based on the evidence, it is clear that the gorilla and the chimp are my first cousins, evolutionarily speaking. Looking through the reinforced glass, the biological similarity is too close to reasonably deny. In claiming that Darwin was wrong about people, I should be clear I am in no doubt that humans have emerged as part of a process of genetic selection. Some humans feel deeply unsettled by their similarity to apes, and seek solace in one version or another of a religious fantasy (Satan, for example, creating the great apes in mockery of God's creation of Man). Others accept the facts of human evolution, but not the causes. The supporters of "intelligent design," for example, do not deny the facts of a 6-million-year-long

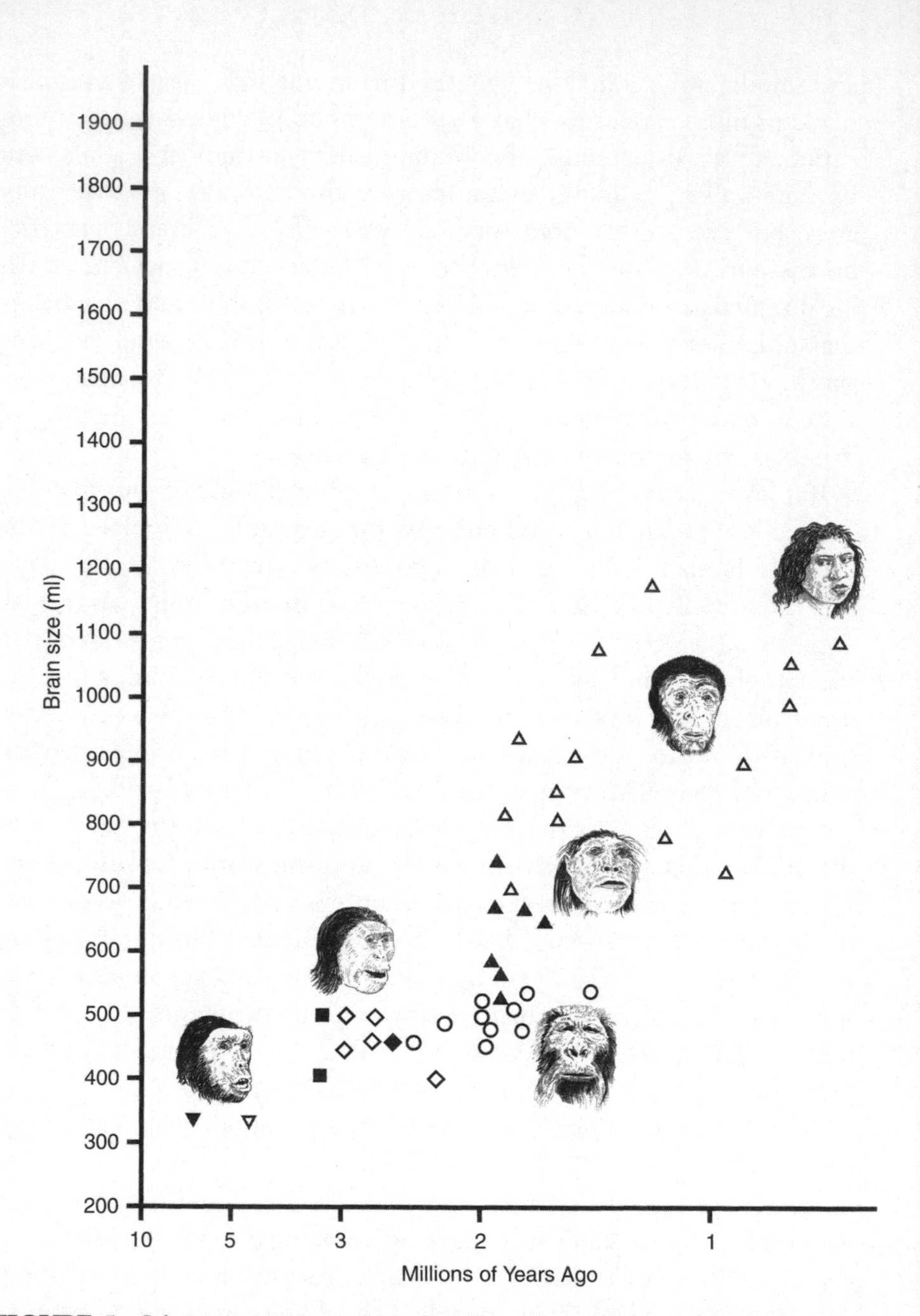

FIGURE 5 Schematic representation of cranial capacity/brain size among hominins (shown on a logarithmic timescale). [▼] *Sahelanthropus tchadensis*. [▽] *Ardipithecus ramidus*. [■] *Australopithecus afarensis*. [◇] *Australopithecus africanus*. [◆] *Australopithecus garhi*. [○] *Paranthropus robustus* and *boisei*. [▲] *Homo habilis* and *rudolfensis*. [△] *Homo ergaster* and *erectus* (including potential late transitional forms).

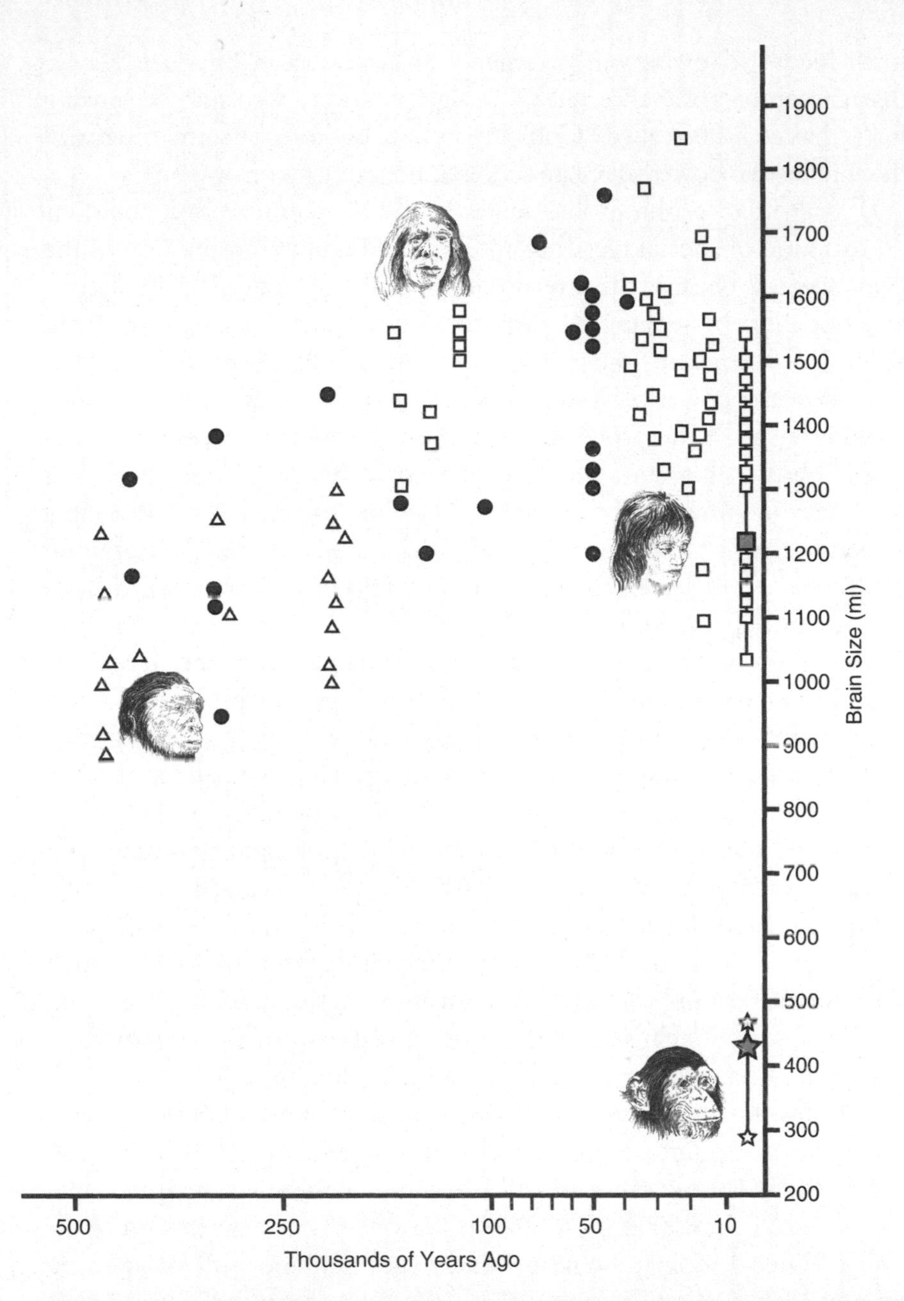

[△] Later *Homo erectus*. [●] *Homo heidelbergensis* and Neanderthals. [□] Archaic and modern *Homo sapiens* (including early transitional forms). [■] Modern human cranial average. [☆] Modern chimpanzee brain size range. [★] Modern chimpanzee cranial average. (Graphic © Frankland/Taylor; primary data from Oppenheimer 2007, with additions.)

fossil record following our divergence from the apes. But they believe that the process was directed by a higher power, a cosmic organizing force that could be called "God" if you like, but at least something with the intellectual power to create successful organic forms.[22]

If Darwinian evolution has never been able to quite shut the door on this kind of special pleading and faith maintained in the face of the evidence, it is because the interpretation of the evidence is, to a degree, questionable. In seeking to dispute some aspects of the Darwinian account as it concerns the emergence of humanity, I hope to be able to show how the power of System 3 has shaped us by significantly changing the terms of the processes of natural selection that Darwin identified. He was not unaware of problems in his own account, and if the scientific movement founded in his name has at times drifted in a direction that is too doctrinaire for its own good, and the greater good of science (as I believe), then it is time to help it re-find its breadth, its balance, and its limitations.

There is a big difference between me and a chimpanzee. It is fashionable to deny it, but I am a lot more intelligent. And I need to be. A chimpanzee could rip me limb from limb with its powerful frame, massive jaws, and huge canine teeth. Facing a chimpanzee in the wild, unarmed, I would be in grave danger. But I am human, and I can protect myself with something I had the brains to obtain, maintain, and operate (such as a rifle with a sedative dart). There are two widely believed reasons for my superiority: the Darwinian view, and the creationist view. Recently there has also been a sort of third way, termed "intelligent design" (or ID). None of them actually make sense, but people go on believing in one or another for want of a more convincing explanation to the puzzle. I aim to provide one here.

My idea is simple. Instead of thinking that human beings evolved from apes to a point where they were intelligent enough to be able to make chipped stone tools and all the other objects we surround ourselves with, I think that these objects created us. I mean this quite literally. When I look at a chimp, what I see is an ape with very strong jaws and large fanglike canine teeth. Because it cannot control fire, and commands only the most rudimentary objects as tools—sticks and pebbles, used as found—the chimp has to eat raw meat and crunch fresh bone. It is very good at this, but it pays a price in terms of biomechanics. It needs a big projecting upper jaw to mount its large canines, and it needs big cheek muscles that attach to the sides and top of its skull to snap its lower jaw shut on its prey with the right amount of force. This means that its skull can only be built in one particular way and there is simply no possibility of its brain expanding. Put bluntly, its jaws pull down against the brain, keeping the skull small and robust. While chimpanzees, from which we diverged some 6 million years ago, all have big canines and a dental diastema, most do not have a sagittal crest (larger males may develop one); by contrast, some later hominins, such as the paranthropines (formally known as robust australopithecines), have a very pronounced sagittal crest. The human cranium is like an eggshell-thin vault in comparison. Thin walled and upwardly domed, it is not held back by any very serious jaw musculature (no chimp would fear my bite).

Despite its genetic similarity to us, a chimpanzee has a small braincase in part because it has to pull down on its own skull to operate its powerful canine teeth. Only after it was possible to soften food by cooking it in a fire, and cut it, not with teeth, but with stone tools, were the biomechanics of the characteristic high-vaulted, thin-domed, human cranium enabled. Fire and stone tools are what allowed us to evolve big brains, and thus become intelligent enough to invent more things. These things, in their turn, had revolutionary effects on our bodies.

In separating fact from fiction, the empirical world is always the best test. Things are most often as they are because that is what works. But when we look at human evolution, we are presented with an untidy collection of partial data—fragments of skull, bits of fossil pelvis, a closely dated specimen from 3.2 million years ago, and another one that might be 4 million years old, give or take a half million years. None of it is flesh and blood, so any ideas about how the various types of hominin ate, slept, reproduced, or defended themselves against attack, have to be inferred.

Practically then, what was the prehistoric equivalent of the security glass that protected me from a nasty attack at the zoo?

The moment when chipped stone tools first appeared was also a moment when the earth's climate tipped from temperate stability into a violent series of ice ages. That unstable period is not yet over. Viewed geologically, it is one big ice age during which glacial advances alternate with rapid retreats. The retreats create brief warm interglacial phases, like the one we currently live in. Our interglacial period, called the Holocene, has been going for some 10,000 years, and is set to continue for at least another 50,000 years.[23]

What is brief for a geologist is not at all brief in terms of the history of civilizations. But we should not think that major climate changes cannot occur on the scale of a human life. The latest high-resolution dates for climate sequences show that at the very end of the last glacial, around 11,000 years ago, temperatures in parts of Britain swung from ice age to as warm (or warmer) than today in a period of just two years.

It cannot, however, be true that the very first chipped stone tool appeared at precisely the same time that the initial big climate shift took place. This is simply because the tool is made during an hour or so as a directed activity while the climate change is a phenomenon that has to be experienced over a somewhat longer period, even when at its most rapid. But that is a rather trivial point and one that should not stop us from imagining an actual causal relationship, with tools appearing as our remote ancestors faced the challenges of a rapidly changing environment.

Understanding timescales and, especially, the technical analytical data by which timescales are established and events positioned in relation to one another, can be hard. To grasp what might have been the case, and to integrate the record of ancestral human activities with climatic shifts, time must be surveyed at once along vast sweeping vistas and in the very thinnest possible slices. A stone pebble can be split in minutes, thrown in a second, cause death in a moment. But when the animal has been butchered, and the blunted, no longer reparable tool has been discarded, the dust settles, the rains come, and sediments slowly cover the scene. A year passes, and then a decade. Undisturbed over hundreds, thousands, hundreds of thousands, and, perhaps eventually, millions of years, geochemical processes turn bone to rock, and encase rock in other rock. Then a moment comes when a human recognizes some kinship in the remains, suddenly exposed after a night of flash floods or a day of hot winds. Brought back into a world of moments and actions, we begin to analyze what once happened. Thus the so-called first family that the human-origins researcher Don Johanson excavated in 1975, the remains of thirteen hominin individuals who he thought must have been caught in a flash flood,[24] are now thought to have been deposited in separate events that caused a more gradual accumulation.[25]

The difficulties of associating a particular tool with a particular cut-marked animal bone, and linking both to a particular hominid fossil, some fragments of jawbone with a tooth or two still intact, have now become extreme. Many of the disputes and disagreements about the origins of humans and the first use of tools result from the fact that fossil remains and stone tools so seldom turn up close enough together for us to be comfortable about inferring an actual association. None of it would stand up in a court of law. But where lots of activity has happened over many decades at some point in the far distant past, the density of finds is enough to show that we most likely do have preserved traces of correlated events and behaviors.

In 1797, in a brickworks near the village of Hoxne in Suffolk in southeast England, some objects came to light that would eventually be seen to have immense significance for understanding the deep human past. John Frere was a man well aware that alternative, more primitive technologies might have existed in the past.[26] A past that was, to him, a far bigger place than many of his Bible-bound contemporaries perceived it to be. Frere was part of the new wave of rational scholarship that was throwing off theological dogmas and superstitious myths.

He knew that women had the same number of ribs as men, that our world was a sphere, not flat, that Jerusalem was not at its center, and it circled the sun, rather than vice versa. He did not believe in the literal reading of the Old Testament, and especially not the strange calculations based on the suspiciously long lives of the patriarchs. According to scripture, Adam lived 930 years, Seth 912, Enos 905, Cainan 910, Mahalalleel 895, and Jared (Adam's great-great-great-grandson) 962 years; finally, in the eighth generation, Methusalah set the record, living 969 years, and dying in the same year—2348 B.C.—that a great flood was supposed to have inundated the earth. Frere rejected the calculation, based on extrapolating back through these implausible genealogies, that heaven and earth were brought into being on the evening preceding Sunday, October 23, 4004 B.C; that Methusalah lived 969 years and died in 2348 B.C., the year of a great, worldwide flood, seemed implausible enough. Frere was certainly not in a position to grasp the processes of organic evolution (for him, humanity was a fixed type). But he could see that it was logically impossible to cite Noah's Flood as the explanation of all the complexities of geological strata.

Frere was a fellow of both the Society of Antiquaries of London and the Royal Society. He, like many learned men of his time, saw that among the apparently tumbled chaos of the world's rocks there was a great deal of stately order, and applied his knowledge of the observable and uniform processes operating in the contemporary natural world to the phenomena of the past. Fossil shells had not been flung halfway up precipitous mountains by massive waves; they existed there in sequential layers, as did the bones of extinct land animals found in marine clays. There was order and gradual progression in the often sedimentary layers that had, it seemed (for the layers were often countable), formed over many millennia. There were caves where stalagmite formations were so large that, unless God had for some reason monkeyed with the basic laws of calcium chemistry, they must have begun to form far earlier than a trivial 6,000 years ago.

The brickworks manager at Hoxne, in the course of digging for clay, had turned up very large numbers of sharp, pointed flint objects, occurring in vast numbers in a layer some six or nine feet down. As Frere writes, "before he was aware of them being objects of curiosity, he had emptied baskets of them into the ruts of the adjoining road." They were found "generally at the rate of five or six in a square yard."[27] The manager may have thought, if he thought about it at all, that they were a kind of petrified icicle, or some other thing of a natural, geological origin (many geological and fossil phenomena have symmetry and standard size). Certainly he did not recognize them as humanly manufactured. There is no reason why he should have done, the artifacts in question having been a contemporary product, some 400,000 years ago.

Frere did not have any means to calculate such a precise chronology either, but he did observe the gradual, multilayered, non-chaotic way in which the geology of the gravel and clay beds had grown. The sharp flints, he wrote, "are, I think, evidently weapons of war, fabricated and used by a people who had not the use of metals.... In the same stratum are frequently found small fragments of wood, very perfect when first dug up, but which soon decompose on being exposed to the air; and in the stratum of sand [which overlay the flints and wood], were found some extraordinary bones, particularly the jaw-bone of enormous size, of some unknown animal, with the teeth remaining in it.... The situation in which these weapons were found may tempt us to refer them to a very remote period indeed; even beyond that of the present world."[28] By the present world he meant the short 6,000-year span allowed by clerical orthodoxy.

Frere was unfamiliar with biological evolution, but he understood that fashions in objects could change dramatically over time. Fast forward to the second half of the twentieth century, and the recognition and recovery of stone artifacts and the independent dating of deposits had progressed so far that it was possible to suggest a detailed evolutionary sequence for technology itself. Five major modes of stone artifact production were proposed[29] (see figures 6 and 7). The earliest and simplest was the breaking up of pebbles to produce an edge. These so-called pebble tool "industries" were identified with the first material culture in the Olduvai Gorge of the East African Rift Valley (and known as Oldowan—this was the sort of technology McGrew considered most characteristic of the Tasmanian Aborigines). Mode 2 showed progress to dual-sided artifacts, so-called bifaces. (First identified at the French site of Saint-Acheul, they are frequently termed Acheulean; Frere's objects were of this type.) Mode 3 represents the more complex "prepare-core" industries of the Mousterian and African Middle Stone Age, and mark a critical step in efficiency of production. Mode 4 is yet more refined, characterized by blades that are often just one part of more complex artifacts in the Upper Palaeolithic period (broadly, the last Ice Age). Finally, Mode 5 labels the ultra-fine "microlithic"

FIGURE 6 Progressive refinement, including increasing symmetry, in stone tool production over time. (Photo courtesy the author.)

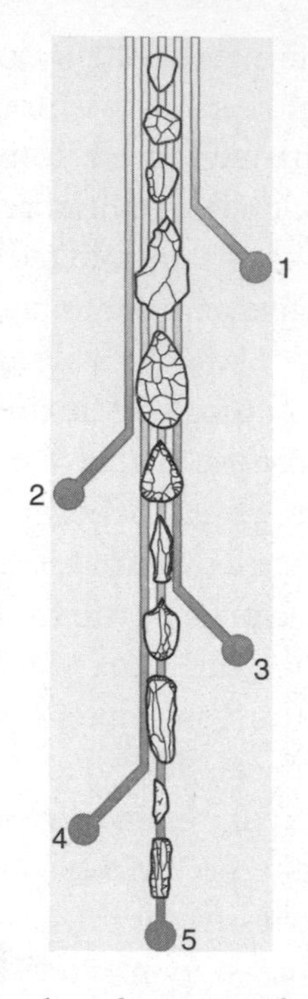

FIGURE 7 Technological modes of stone artifact production: Mode 1. Pebble tool industries (Oldowan); Mode 2. Biface industries (Acheulean); Mode 3. Prepared-core industries (Mousterian, Middle Stone Age); Mode 4. Blade industries (Upper Palaeolithic); Mode 5. Microlithic industries (Mesolithic). (Graphic © Frankland.)

technology of the Mesolithic hunters who existed prior to the advent of farming (microliths were often used together, set in rows to make a continuous cutting edge, each element being replaced as it wore out).

The first definitive Mode 1 objects we know of, a couple of thousand deliberately shaped flakes of stone and pebble cores, found at the site of Gona in Ethiopia, could be as young as 2.52 million years old, or as old as 2.60 million.[30] An 80,000-year zone of uncertainty is not too bad this far back; what is certain is that the period when they were made was a time of climatic upheaval. It is tempting to see the two as related and, in turn, connected to the forces that powered human evolutionary emergence. Perhaps environmental challenges stimulated or spurred intelligence; perhaps the first artificial technology was an adaptive response. If that seems too neat, then it probably is. Under the magnifying glass things do not match up as we might like, and the order of events shows up as subtly, yet fundamentally, out of synch.

We are much more certain about the way the climate changed than we are about hominin evolution or the birth of technology. Climate is part of nature, and its changes have a consistent and global impact. Levels of oxygen isotopes in deep-sea cores and the paleomagnetic records from the great banks of accumulated loess soil in China can be correlated at a high level of resolution to produce a very accurate picture of changes over time.

In a general sense, it is correct that humans emerged at a time of climatic cooling. Previously the globe had been warm and humid, grading perhaps to temperate in places. This was the long, stable Tertiary geological epoch that had begun 65 million years ago with the extinction of the dinosaurs and that had witnessed the emergence of mammals. India's collision with Eurasia had slowly forced the Himalayas up, altering the circulation of air in the Northern Hemisphere by degrees. Certainly, cooling had been slow and steady for a long time before the icy crunch came. Northern Hemisphere cooling becomes a trend, discernible from around the start of the Pliocene epoch, beginning a little before 5 million years ago (5 ma). It was deepening by around 3.6 million years ago, when winter ice started to hang on into the summer months. In the south, ice had by now long covered the Antarctic continent, which had been steadily drifting toward its present extreme position after the dinosaurs had gone. Now, in the far north, year-on-year ice buildup commenced, and glaciers took shape there too. This created a vicious cycle that helped trigger a much sharper episode of cooling.

Around 2.72 million years ago the summer monsoons weakened dramatically. A major expansion of ice sheets followed at 2.60ma—a freezing pulse that was followed by another at 2.52ma, and then two more, at 2.49ma and 2.44ma. As these four strikes rang like a death knell, the earth formally entered the Ice Age. Both planetary poles were now simultaneously iced up, perhaps an unprecedented event in global history.

A key date recently fixed by the International Commission on Stratigraphy is 2,588,000 years ago, defined by a moment of magnetic reversal known as the Gauss-Matuyama. This provides a key orientation point for geologists and paleoclimatologists, the point when the Quaternary period might be said to begin.[31] We still live in the latter part of it, the Holocene epoch, which has followed the Pleistocene, and during which there has been a very dramatic recession of glaciers and the pack ice at the North Pole has grown ever thinner. But we should not count on having escaped the sequence of big chills. We remain at the mercy of a system in which glacial re-advances (an "ice age" as a noticeable event) can be expected to occur, if not like clockwork, then with regularity. The way our planet rotates is unsteady, and as we wobble, the way we catch the sun's heat varies. A main oscillation of this Milankovitch cycle occurs over a period of 41,000 years and corresponds fairly neatly with the sequence of ice pulses until around a million years ago. But there are other issues, such as the changing overall shape of the earth's elliptical orbit as it is pulled by other planets. The complex interaction of the earth's movement set up resonances not just at 41,000-year intervals, but at 100,000-year and 400,000-year intervals. For reasons that are still not entirely understood, the ice pulses shifted around a million years ago to follow the 100,000-year cycle.[32]

Around 110,000 individual annual layers are actually countable by eye when a deep core is taken out of the Greenland ice (the sort of uniformitarian data that biblical creationists have to become especially imaginative in order to deny); the very bottom, compressed and distorted, may preserve ice dating back to the original Ice Age onset 2.6 million years ago. Another raft of glaciers is expected sometime in the future. Most climatologists expect their arrival in rather more than 50,000 years time, while a few believe that the recent artificial boost to warmth-trapping atmospheric carbon dioxide may have altered the system entirely, and the ice may never come back. In a dramatically opposite scenario, it is also just possible that the human havoc in the current climatic system could lead us into a period of far greater instability and actually bring on a glacial period more quickly. Whenever it comes, our current bout of global warming may turn out to be no more significant than the transitory heat generated by a child wetting its pants as a blizzard starts coming in (and if we are responsible for the warming, then perhaps it is just as ill-judged, serving only to increase discomfort in advance of a more severe survival challenge that lies some way ahead).

In really big-scale geology, the start of the Quaternary period, with its alternating glacial freezes and interglacial thaws, is very recent. If we imagine planet Earth as a year old, from formation to now, then the Quaternary started on the last day of the year, the 31st of December, at around 7 o'clock in the evening. Life may have been around quite early, but did not properly get cracking until the last two months (the Cambrian); dinosaurs started to really strut in December, but got wiped out by around Christmas Day, perhaps in connection with an asteroid impact (roast Tyrannosaurus might have tasted similar to turkey). The scene was now clear for mammals, which had been kept down by the scary old regime, to thrive. On this timescale, modern humans appear well within the last half hour, barely time to get the New Year's champagne chilled down (wine actually appeared only in the last half minute, and glasses to drink from in the last ten seconds).[33]

Leading our individual lives, this compression of timescale is mind-boggling. There is no real way of grasping the scale and the resolution needed when attempting to argue cause and effect without some sort of relativizing. The dinosaurs went extinct 65 million years ago, and mammals began their ascent. Sometime after 23 million years ago, the first apes appeared, first in Africa and spreading out by 16 million years ago. The period—the Miocene epoch—had started out very warm, with the ice over Antarctica melting, and this allowed the rapid and complex diversification of monkeys and apes throughout the luxuriant tropical forests. But, in the later Miocene, the Sahara began to go from moist and fertile plain to arid desert.[34] Although itself a hot place, its effect was to dry and also to cool parts of Africa and Eurasia, while setting up a massive ecological boundary. From around 10 million years ago onward, suitable habitats for Miocene apes became rarer, and different species were forced into greater territorial competition with one another as rainforests shrank.[35]

Between 3 and 2 million years ago is a frustratingly quiet period in the fossil record. Either not many bones got well fossilized (perhaps because of the climatic upheavals) or we have been looking for them in the wrong places.[36] It is hard, therefore, to decide whether the emergence of our genus (*Homo*, to which our species, *sapiens*, belongs) was an environmentally driven event. Coincidence is not cause, and other factors were probably in play. A complicating factor that at least some paleoanthropologists seem keen to skate over is that there is no fossil evidence for genus *Homo* at the point when stone tools first appear at sites in Ethiopia that seem clearly associated with the little *Australopithecus garhi*, with its essentially chimp-size brain.[37] The very earliest species of upright walking ape accepted (though not by all) as belonging to our human genus is *Homo habilis*. But this creature cannot be documented until 2.33 million years ago.[38]

Culture is less predictable, hard to be sure about. Artifacts produced by early cultural systems do not always survive. When they do, their presence on a few special sites, mainly in Africa, may not be representative. The objects found at Gona are of different kinds, and there are a lot of them. This was not some soon-to-be-aborted experiment. What is revealed is a vibrant, well-established toolmaking tradition. It must have been preceded by something more experimental and less certain.

Working with decimal points and attempting to gauge the precise age of fossils and particular stone tools this far back in time is fraught with technical challenges. New data are continually being analyzed, and refinements in the techniques used for measuring ages and reducing the errors that can be associated with such measurements are in more or less continuous development. So, for the time being at least, there is a gap between the date for the first chipped stone tool (no more recent than 2,520,000 years ago) and the earliest definitive date for a species to which the genus name *Homo* has been applied by some (not all) paleoanthropologists (no earlier than 2,330,000 years ago).[39] I have put in all the zeros so that the actual time gap, some 190,000 years, becomes clear. To set that in perspective, 190,000 years is pretty close to the total length of time that our own species of human, *Homo sapiens*, has been on the planet. It is not a trivial time span. What is its significance?

CHAPTER 4

THE 7,000-CALORIE LUNCH

I picked up a small dying wallaby, whose mother had thrown it from her pouch. It weighed about two ounces, and was scarcely furnished yet with fur. The instant I saw it, like an eagle I pounced upon it and ate it raw, dying as it was, fur, skin and all. The delicious taste of the creature I shall never forget.

—Ernest Giles, *Geographic Travels in Central Australia* (1875)[1]

IF PUSHED TO THE LIMIT, we can eat almost anything. The body knows what it needs, as did Ernest Giles, returning from his first unsuccessful attempt to cross what would become known as the Gibson Desert. He and his co-explorer, Alfred Gibson, had set out from Fort McKellar on the northern flanks of the Rawlinson Range some 150 miles west of Uluru (Ayers Rock) to explore the wilderness but ran out of water.[2] Returning in advance of Giles, Gibson somehow never made it to the place they knew as Circus-water, perishing somewhere in the blazing heat. But this account is nevertheless shocking, not just for the privations implied, but because it breaks two taboos at once. First, with rare exceptions, we believe that we should eat animal food after it has been killed in a separate process. And second, that process should be followed by the one known as cooking.

There was a certain poetic justice in operation when I failed to find the biological anthropologist Richard Wrangham's book *Catching*

Fire: How Cooking Made Us Human on the otherwise admirably well-stocked popular science shelves of my local bookstore.[3] Wrangham, of Harvard University, has produced a series of scholarly papers over the past decade in which he argues that human evolution, especially our progressive, unprecedented increase in brain power, could not have happened in the absence of regular supplies of cooked food, especially cooked animal protein.[4] Control of fire and the development of social relations around cooking were what, according to this attractive theory, allowed us to stop being chimps. Keen to read Wrangham's argument complete, I asked the staff to check availability. It was in stock, the computer said. But no one could find it. At least, not until someone looked on the cookbook shelves, and there it was, dust jacket emblazoned with the statement "Once our ancestors began cooking their food, the human digestive tract began to shrink and the brain to grow."[5]

In the annals of humanity's big accomplishments, cuisine has been underrated. Appreciated of course, but never quite understood as an established high art, an apex of technology. Why does the chocolate cake made at the Café Sacher in Vienna not figure alongside Michelangelo's *David* as a miracle of art and technology? The secret recipe is a wonder and a puzzle: Is there a hint of black pepper? blade mace? And how is the micron-exact icing achieved? Generic *Sachertorte*, available at a fraction of the cost in *Konditoreien* across Austria, can be good in its own way, but at worst it stands in relation to the real thing in the same way that *David*-based molded garden art stands to the Renaissance sculptor's magnificent original marble sling-thrower.

Cake making in the Austro-Hungarian tradition entailed as many complex chains of supply, quality assurance, craft skill, and artistic inspiration as the construction of the elaborate urban architectural setting in which its products were consumed (and to which they were aesthetically related: "more like pastries than buildings," as the Russian writer Anton Chekhov commented, impressed by both on a visit in 1891).[6] Yet baking cake lacks status as a serious cultural achievement. Perhaps this is because it is often seen as women's work, or, at least, as belonging to the realm of domestic production, and it has too often been men who have decided what is really important in species history. Maybe it is underrated because it is impermanent, fugitive, vanishing. Yet Mozart's music fades away at the end of every performance; and recipes, like composer's scores, survive. To make it a reality, the technology has to be maintained, whether it involves cake tins or pianos, and the personnel have to be trained, and dedicated to producing versions for the overlapping cognoscenti of café and concert society down the generations, long after the first performance, centuries after the artist-inventor has died (Wolfgang Amadeus Mozart died in 1791, Franz Sacher in 1907).

That cooking is a deeply cultural practice and one of the defining characteristics of our species has, of course, long been recognized. The French essayist and thinker Michel de Montaigne maintained that the art of dining well was "no slight art,"[7] and in the 1770s, Dr. Johnson's biographer, James Boswell, elaborated the sentiment into a species definition. During dinner table conversation with the political philosopher Edmund Burke, Boswell averred that Benjamin Franklin's definition of man as a toolmaker (*Homo faber*) was flawed, as not everyone made such things: "My definition of *Man* is, 'a cooking animal.' The beasts have memory, judgement, and all the faculties and passions of our mind, in a certain degree; but no beast is a cook."[8] Some two centuries later, one of the foundational texts of sociocultural anthropology, Claude Lévi-Strauss's *The Raw and the Cooked*, argued that our own definition of ourselves as not animals is consistently related to the technology and rituals of food preparation: nature and the wild is "raw," culture and the tame is "cooked."[9]

Lévi-Strauss argued that what foods people eat, and when and how they eat them were truly central issues, bearing on life and death. Eating the wrong food can lead to death, but not just because it is physiologically wrong. It can as easily be—indeed more often is—culturally wrong. It is sometimes hard for us to grasp this, living in a world of (it seems at least) near-constant twenty-four-hour consumer supply. We may have personal tastes and foibles, and we are aware of the dietary interdictions of different religious and cultural groups around us. We know it can get serious. Nevertheless, it is still a shock to hear a particular ethnographic description of the fate of an Eskimo girl in coastal Labrador who persisted in eating caribou meat in winter, despite caribou's being an exclusively summer food. The anthropologist Mary Douglas wrote that it was a "trivial breach of an abstinence rule, it seems to us. But by a unanimous verdict, she was banished in midwinter.... These Eskimo have constructed a society whose fundamental category is the distinction of the two seasons. People born in winter are distinguished from those born in summer. Each of the two seasons has a special kind of domestic arrangement, a special seasonal economy, a separate legal practice, almost a distinct religion."[10]

Just as winter hunting gear must never come into contact with tents used in summer, so summer fish, such as salmon, if dried and preserved, cannot be eaten in winter as it will come into contact with winter food in the person's stomach. Douglas judges that "By disregarding these distinctive categories the girl was committing a wrong against the social system in its fundamental form. A lack of seriousness about the categories of thought was not the reason given for why she was condemned to die by freezing. She had to die because she had committed a dangerous pollution which set everyone's livelihood at risk."[11] The analysis of what happened in this case seems to have very little to do with the terms of bodily survival, of actual digestive biology. In fact, with enough caribou meat in some stash somewhere, the girl could perhaps have physically weathered the worst of the cold, despite transgressing the metaphysics of her culture. But it is not clear that things can be separated out like this. The cruel verdict, justified in the terms above, must also have been meant as exemplary, enforcing a more general cultural conformity through fear. Non-trangressible norms help maintain a routine for living, for maintaining an ongoing human community in an extreme environment. The means being tried and tested, the most minor dissent in such societies may lead to severe punishment, a drift from "the way of our people" presaging chaos and the destruction of all.

The Labrador Eskimo manner is not the only way of surviving in a demanding high-latitude habitat. Other cold-adapted peoples have very few cultural rules governing everyday food consumption. In 1820, the British naval officer John Cochrane set out to walk from the French coast at Dieppe to Saint Petersburg and then on, right through Russia, to the Sea of Okhotsk and the Kamchatka peninsula.[12] Traveling largely without funds so he would not be robbed too often, he had letters of introduction to various officials, who helped him to stay clothed and provisioned in a periodic way, yet he often survived only on the spontaneous hospitality of the tribespeople he encountered. Beyond the Urals, he found himself among the Tungusians and the Yakut of the central Siberian plateau. Winter drew on, and the temperature dropped to minus 30° Réaumur (–35° Fahrenheit or –37° Celsius). That was in the daytime. At night, Cochrane slept in uninhabited "charity"

yurts, or in the temporary company of Yakut trackers, braving the elements outdoors, warmed only by special horseshoe-shaped all-night campfires.

The Yakut were people who had the habit of "eating whenever there is food, and never permitting anything that can be eaten to be lost." In the town of Tabalak, Cochrane watched mesmerized as three of his tallow candles, "several pounds of sour butter," and a large piece of yellow soap were all eaten by a voraciously hungry small boy. He reflected that "there is nothing in the way of fish or meat, from whatever animal, however putrid and unwholesome, but they will devour with impunity, and the quantity only varies from what they have, to what they can get."[13]

Cochrane was a stickler for quantification and amateur statistics and was well versed in the rations for conversion between Russian and British systems of measurement. Clearly he is serious when he writes (although not expecting the reader will find it easy to believe), "I have regularly seen a Yakut or a Toungouse devour forty pounds of meat in a day. The effect is very observable upon them, for, from thin and meagre-looking men, they will become perfectly pot bellied." Cochrane supposes that "Their stomachs must be differently formed from ours" before proceeding to provide accounts of varying pedigree concerning daily food intake among the Yakut, such as (for one man) the hindquarter of a large ox, twenty pounds of fat, and a proportionate quantity of melted butter in drinks.[14] Finally, he records a test made by a Russian admiral named Saritcheff. Impressed by tales of massive appetites, but skeptical, Saritcheff made up a rice porridge for a particular Yakut. The porridge, with its three pounds of butter added, weighed in at twenty-eight pounds, "and although the glutton had *already breakfasted*, yet did he sit down to it with the greatest eagerness, and consumed the whole without stirring from the spot; and, except for the fact that his stomach betrayed more than ordinary fullness, he showed no sign of inconvenience or injury, but would have been ready to renew his gluttony the following day."[15]

These massive amounts may seem amazing to many of us whose energy needs are compensated for by habitually warm, dry clothing, centrally heated environments, and sedentary activities. But we do not have to go very far back in Western culture to find different attitudes, perhaps even different aptitudes. In 1872, the Englishman George Musgrave was enjoying the cuisine of Normandy and described in some detail a single lunch being eaten by a honeymooning couple on board a river steamer at Rouen. As summarized by the cookbook writer Elizabeth David, it consisted of "soup, fried mackerel, beefsteak, French beans and fried potatoes, an omelette *fines herbes*, a *fricandeau* of veal with sorrel, a roast chicken garnished with mushrooms, a hock of ham served upon spinach. There followed an apricot tart, three custards and an endive salad, which were the precursors to a small roast leg of lamb, with chopped onion and nutmeg sprinkled upon it. Then came coffee and two glasses of absinthe, and *eau dorée* [Goldwasser, another spirit], a Mignon cheese, pears, plums, grapes and cakes." In the drinks department, "Two bottles of Burgundy and one of Chablis were emptied between eleven and one o'clock."[16]

This is a significant meal. On honeymoon, perhaps the calories need replenishing more often. But how many calories in this meal? Nutritionists suggest that a normal healthy intake of calories is in the region of 2,000–2,500 per day. Calculating the calories in this meal, in ballpark numbers, was by no means easy. Standard calorie counting websites provide conversions for every unpleasant convenience food under the sun, containing assessments for slices of "Turkey-lite" rather than apparently obscure food items like an omelette (unrecognized by Google's calorie calculator).

Allowing that we do not know how big the eggs were, how much butter was in the omelette, whether the happy couple shared a large one and so on, the average chicken egg contains about 80 calories. In the regional style in question, according to my *Recettes de Normandie* of 2001, the whites may be separated first to whisk into the yolks, and eight eggs are used to serve four people as an appetizer.[17] Assume 160 calories each, excluding the butter.

Let's total up the calories. To the 160 calories of the omelette, let's add soup (a standard serving could be 100 calories, but Normandy recipes are hearty, so 150 calories), fried mackerel (250 calories would be a medium fillet—I suspect these were whole, but small), steak (800 calories), beans (150 calories), fried potatoes (400 calories), half the roast chicken (750 calories minimum), half the ham hock and the small leg of lamb (1,000 + 600 calories), apricot tart (let's make it small—400 calories), half a cheese (conservatively 250 calories), cakes (250 calories), and the (presumably little) set custards (made with eggs so in the same region as the omelette = 160 calories). A bottle of wine is 650 calories, and spirits 50 calories per shot, so per person we have alcohol at 1,100 calories. I haven't rated the endive salad. Did they have sugar in their coffee? Who cares. That comes to 6,420 calories, which does not include any of the fruit, nor the butter, cream, and oil in the cooking and the dressings. It is not unreasonable to suggest around 7,000 calories per person. The estimate might be out by a thousand calories either way, but we must be in the right zone for this romantic riverine lunch.

The honeymooners obviously enjoyed their food as much as children do when it is plentiful and the rules are relaxed (childhood obesity experts Carnegie Weight Management calculated in advance of Christmas Day 2009 that British children might well be able to get through 6,000 calories from dawn to dusk).[18] The French couple were, however, adult, and possibly more attuned to large-scale eating over a mere two-hour period. We are not to imagine that they had forgone breakfast, and they probably looked forward to supper as well, if not perhaps any afternoon tea in between.

Modern nutritionists are easily scared by such amounts. In the United Kingdom, the Committee on Medical Aspects of Food Policy (COMA) in 1991 issued guideline daily intake amounts of 2,000 calories for women and 2,500 calories for men. In 2009 the Scientific Advisory Committee on Nutrition decided that COMA had underestimated average physical activity levels, and thus the rate at which food is burned for energy, and so relaxed these figures by an average 400 calories. Activity is the critical extra factor, of course. If you take a lot of exercise, then you need more fuel. Army ration packs worldwide typically range between 3,000 and 4,000 calories a day, and they represent a baseline—you eat more if you can get it. Actually, according to the measurements of sports physiologists, if the honeymoon couple had spent their afternoon playing professional ice hockey, they could have burned off a full 6,000 calories and lost around fifteen pounds of body weight into the bargain; distance swimming burns off similar amounts (which is why Olympic swimmer Michael Phelps claims to eat 12,000 calories a day[19]).

Cochrane's claim of forty pounds of meat a day for the Yakut, which he says he saw regularly, does not now look quite so implausible. Meat varies greatly in caloric value according to how fatty it is. A pound of lean game comes in at around 600 calories, with prime steak and belly pork rising to 1,500 or even 2,000 calories. Lard and butter are around 3,800 calories per pound. Assigning a rather lean 1,000 calories per pound to the Yakut meat cuisine, we have 40,000 calories in a day. That's a whole lot of ice hockey. But in central Siberia in winter, with temperatures so deep in the minus numbers, that might be burned off in a couple of days of tramping. In heavy snow it might be the time it takes to manage the twenty-five miles between charity yurts. Or not: Cochrane, who clearly had immense reserves of stamina and resolution, bemoans the fact that this spacing was just too far in bad weather and rather too short on clear days (which is why he sometimes ended up sleeping out). But many other accounts confirm that these Siberian folk ate as much as they could when they could because they also had, all too frequently, to put up with near starvation. Cochrane is surely right to suppose that their metabolisms had adapted to this on-off pattern.

To see what this may tell us about prehistory, and human evolution, we must turn to Richard Wrangham's elegant theory, his big evolutionary idea that not only are humans obsessed with cooking at a symbolic level, as social anthropologists like Lévi-Strauss and Mary Douglas long ago established, but they are also biologically dependent on it. On the details where I have an expertise to judge, Wrangham passes muster, and I accept much of his argument entirely. What has been new to me has made sense. But (there is a but), I am not persuaded that cooking in and of itself provides the whole explanation for human brain-size expansion. I am not sure that Wrangham really claims that it does. And he probably takes his argument as far as it can safely go on the basis of the available evidence. Archaeological theoreticians can make fools of themselves where primatologists cannot. My job then, after introducing the main points of Wrangham's *Catching Fire* theory as fairly as I can (and as briefly, duplicating a great read being pointless), will be to take us further, with less certainty.

Wrangham sees cooking as an absolute necessity for humans, compensating for a biological deficit that technology—in this case primarily the control of fire—has caused in our species. He argues persuasively that Claude Lévi-Strauss got one key thing entirely wrong: humans did not cook in order to differentiate themselves from animals. It was not a *chosen* badge of culture. If it had been, cultures being what they are, some peoples would have prided themselves on being raw foodists (of which more shortly). Instead, rather more logically, the reason for the ubiquity of fire and cooking in every culture of the world is because it is essential to our species' biology. We need to do it, so we do do it, and so we are in a position to notice that animals (quite typically the things we cook) do not do it. From that arises the symbolic dimension, which is, in this crucial sense, a secondary feature.

Our digestive system was dramatically compromised by the emergence of upright walking. The change in posture resulted in intestines far shorter than those of our fellow apes. Our guts, as calculated by Leslie C. Aiello, president of the Wenner-Gren Foundation for Anthropological Research, are a mere 60 percent of the size that they should be, when compared to those of other primates.[20] We have a sharply reduced internal surface area for food breakdown and absorption. A 450-pound mountain gorilla can live quite happily on a diet of raw leaves and fruit, but it is impossible for a human to exist on the same diet. The issue is not just a lack of high-quality protein from meat. It is questionable whether any human group has survived for any length of time on a wholly raw, wholly unprocessed diet. This is because our gut-length deficit, bad enough if we only had gorilla-sized brains to maintain, is exacerbated by the massive energy requirements of our enlarged brains. The mismatch is so great that Richard Wrangham argues that the abilities to hunt (to get large amounts of high-energy, high-protein meat) *and* to make fire (to cook the meat into a more digestible form, effectively adding energy from burning wood into the total equation so that our internal digestion can function more cheaply) were equally critical to the emergence of human intelligence.[21]

Because the energy equations of a short gut and a large brain simply do not work out, if you want to pretend to be adapted to salads, raw fish and meat, and uncooked vegetables, then you need to play a different technological trick. If you live with access to warm clothes, central heating, food-processors and liquidizers, and protection from the elements, and do not attempt anything too strenuous, you can just about manage. You might say it is living raw, but in reality you are adding a lot of additional, usually fossil-fuel-based, energy to your total system. It's energy that could probably be more efficiently put directly into your food, by cooking it.

Despite our nutritional disadvantage, or in fact, because of it, and through it, we have developed a technology of cooking and hunting that allows us, uniquely among apes, to inhabit almost every environment on the planet. Or, put another way, we use technology to create homeostasis: to equalize all environments to one, and to maintain energy intake and body temperature whatever the outside conditions are. It may very well be that we have been so successful as a species because of our "phenotypic plasticity"—the so-called Baldwin effect.[22]

What this means is that we are able to develop according to our environmental conditions along a number of different pathways from birth to adulthood. Inuit and other northern-latitude marine-mammal hunters grow up to cope with levels of fatty acids that would prove lethal not just to someone unused to their diet, but to anyone not brought up among them from birth.[23]

The diets of hunter-gatherers may seem strange: Chekhov writes, for instance, that

> The Gilyak has a strong, stocky build, and he is of medium or short stature. Height would be of no advantage to him in the taiga. His bones are thick and distinguished by the strong development of his limbs [which] indicate powerful musculature and a perpetual, intense struggle against nature. His body is lean, without fat. There are no stout or corpulent Gilyaks. All his fat is used for the warmth which a man on Sakhalin must generate in his body in order to compensate for the heat loss caused by the low temperature and the excessive humidity. It is understandable that a Gilyak should need a good deal of fat in his diet. He eats fatty seal meat, salmon, sturgeon and whale fat. He also eats rare meat in large quantities in raw, dry and frozen form... the Gilyaks consider agriculture a grievous transgression; whoever ploughs the land or plants anything will soon die.[24]

A flexible developmental biology, coupled with a technological culture, can make humans extraordinarily adaptive to different environments: the tools equalizing and homogenizing conditions as far as possible, and the individual biology making up the deficit. The realm of material artifacts—System 3—thus crosscuts the survival of individual genes, or indeed individual organisms. The survival of the fittest does not apply to human techno-cultures so much at a personal level: instead, cultural adaptation through the deployment of smart technology creates conditions suitable for entire communities to succeed in.

We need food, compete for it, and may even recognize a source in each other. Lynn Kilgore's study of trauma on chimpanzee skeletons includes work at the famous Gombe reserve, where the life cycle of the generations has been followed via detailed documentation of individual apes from birth to death. One female, named Gilka, had been a particular target for aggression from another female and her daughter. All three of Gilka's offspring, born over a seven-year period, were removed by this pair, two definitively eaten, one never found. Gilka, attempting a revenge attack, was bitten in the hand and died of chronic infection. Aggressive cannibalism has been well documented among the Gombe chimps, and although some doubts have arisen as to the validity of observations made on a community that is being provisioned in part by a research project, the logic is a natural, Darwinian one.[25] Cannibalism is widespread in nature, with approaching 1,500 species implicated. Some cannibalize unwittingly, especially the extreme "r-strategists." These are creatures, like many fish and insects, that have vast numbers of offspring and make no parental investment in them: the young take their chances equally to become prey or the next successful generation. But among mammals, where over seventy-five species are now known to cannibalize, there is a more calculating approach.

Mammals eat their own young in some circumstances. I will not quickly forget the deep shock of my daughters, who were discreetly watching the magic of a friend's cat giving birth to an overly numerous litter only for the mother to start swallowing her young back down, even biting the head off one of the kittens. The mother cat ended up with a suitable number to suckle, her energy levels replenished by the ingestion of the ones she had ruthlessly and logically singled out as weak. An equally well-documented mammalian approach, seen among lions and, disturbingly, among chimpanzees, is the eating of a rival's young. There is a dual survival logic in this, taking competitor genes out of the equation while boosting one's own nutritional health, and therefore one's potential reproductive success.

Cannibalism as the end result of conflict occurs among humans too, although the data may be hard to assess. John Tanner—who, at the age of ten, in around 1790, was taken as a slave by the Shawnee tribe of American Indians and thence sold to the Ojibwa, among whom he became a naturalized Saulteaux speaker and a married man—incorporated a chilling little diagrammatic drawing toward the end of his celebrated *Narrative of a Captivity*, published in 1830.[26] It is a mnemonic image for a "War Medicine Song": a triangle containing a stick figure with a line descending down to it and a circular object at eye level. Tanner writes:

> This figure, the words for which are lost, or purposely withheld, represents a lodge, a kettle, and a boy, who is a prisoner. The line from

> his heart to the kettle, indicates too plainly the meaning of the song. I know not whether any still doubt that the North American Indians are cannibals; if so, they are only those who have taken little pains to be correctly informed. The author of the preceding narrative had spent the best years of his life among the Ojibbeways; a woman of that tribe was, as he somewhere says "the mother of his children"; and we need not wonder that, after becoming aware of the strong feeling of white men on this subject, he should be reluctant of speaking of it. Yet he makes no hesitation in saying, that the Sioux eat their enemies, and he once admitted, that in the large Ottawwaw settlement of Waw-gun-uk-ke-zie, he believed there were few, if any, persons living in the late war, who did not, at some time or other, eat the flesh of some people belonging to the United States.[27]

But the concept of cannibalism is hard to accept in the sheltered conditions of the modern consumer world, and a paucity of modern data, coupled with a revisionist form of social anthropology that has viewed cannibal claims as the racist and self-justifying slanders of Victorian explorers and missionaries, meant that by the 1980s the subject—along with ritual human sacrifice, child slavery, genital mutilation, and a number of other uncomfortable indigenous practices—had been all but expunged from student textbooks. Eyewitness accounts, such as that of Captain James Cook among the Maori on Fiji, were dismissed as implausible, for no good reason except that the behavior he described—the ceremonial eating of enemies—is not a tradition that has been carried on in the developed world. Cook twice permitted human flesh to be cooked on the quarterdeck of the *Resolution*, and he signed a round-robin affidavit that it had happened in his presence. But his view was biased by the generous Enlightenment idea of "the noble savage." He saw the activity as an understandable overflow of a heroic temperament that had been victorious on the field of battle, and thought it happened only in that context. In this he was contradicted by the expedition astronomer, William Wales, who said that cannibalism was also practiced from choice, "in cool Blood": "It cannot be for want of Animal food; because they everyday caught as much fish as served both themselves and us: they have moreover plenty of fine Dogs which they were at the time selling us for mere trifles."[28]

One of the things that has stood in the way of our appreciation of the existence of cannibalism has been the (biologically correct) idea that we are all one species. We may fail to understand that, for particular human cultures, humanity extends to their borders and no farther.

Those creatures in the next valley or on the next island are considered animals, and therefore appropriate targets for predation. And if, as seems likely, *Homo sapiens* emerged from an evolutionary period of intense survival competition between different species of upright-walking ape, then the mental faculties for identifying otherness and difference may be both finely tuned and indelible.[29]

There are currently at least twenty-six different published definitions of what a species is, and how it differs from a subspecies. The way that biological taxonomists break up and classify the living world is hugely complex. The split into kingdom, phylum, class, family, genus, species, and on down to subspecies, breed, race, or strain, is based on principles of grouping creatures together on the basis of their most basic shared characteristics, such as having an internal skeleton or an external shell, and working down to the level of greatest similarity, usually defined as two creatures successfully able to breed with one another. Modern humans can be defined as *Animalia* (the kingdom: including most things with a nervous system); *Chordata* (the phylum: all creatures that have a spinal chord); *Mammalia* (the class: warm blooded, giving live birth to young who suckle); *Primates* (the order: this one is rather debatable, due to a controversy over the importance of the difference between wet and dry noses, but if considered as a single group, it currently includes lemurs, tarsiers, marmosets, and all monkeys and apes[30]); *Hominoidea* (the super-family: including all apes and humans; *Hominidae* (our family: the great apes and us); *Homininae* (our sub-family: chimpanzees and us); *Hominini* (our tribe, but not in a cultural "tribal" sense: we are the only known living member, but all our extinct upright-walking ancestors and also-rans belong to it and are thus known as hominins); *Homo* (our genus: excluding australopthecines, paranthropines, and other prehistoric bipedal apes); plus *sapiens* (our species: we belong to this wherever on the planet we were born and whatever we look like).

In reality, only individual creatures exist, and whether they personally perpetuate their own species or not is the only thing that really matters as far as evolution is concerned. The ever more generalizing and inclusive taxonomic levels above the species level, while they have to be rather arbitrarily decided, have been developed so as to reveal the best current understanding of the deep-time historical relationships between related groups of species as they have fissioned off from one another.

A species means a type, individual members of which can and do breed with one another, while closely related species, which are said to belong to the same genus, share a common ancestor, not far back in evolutionary time, but have grown apart due to geographical separation or evolutionary specialization, so that they no longer meet to breed, and if they do, produce hybrids that are typically sterile.

It is clear that hominins began to emerge as early as 5 million years ago, and our genus, *Homo*, became distinct around 2 million years ago. Our sort of human—"anatomically modern humans"—was present at least 150,000 years ago.[31] Against this deep history, the Aboriginal colonization of Australia and Tasmania is a recent event, occurring within the last 40,000 years. Aborigines are (and were) modern *Homo sapiens*. And modern *Homo sapiens* is a single species in which innate intelligence varies more among the members of any given group than between groups. We are a species in which normal healthy reproduction is limited by choice, opportunity, society, and culture, not by biology.

Darwin was drawn to the idea that humanity was composed of several separate species in conflict ("The varieties of man seem to act on each other in the same way as different species of animals—the stronger always extirpating the weaker"[32]). However, he knew well that European sailors had interbred with native people whenever the opportunity had arisen, so he opted for the view that human races were more like *sub*species, between whom hybridization or interbreeding could occur, as it did elsewhere in nature. It was a close call. The sentence that appears as an epigraph to chapter 2, "I could not have believed how wide was the difference between savage and civilised man: it is greater than between a wild and domesticated animal," carries the clear implication that Darwin judged interracial marriage closer to "hybridization" (think donkeys with horses) than to "interbreeding" (dogs with wolves).[33]

Darwin regarded the Aboriginal Australians as a subspecies perceptibly closer to the apes than white Europeans were and (as we see later) took a dim view of any dilution of superior grades of humanity by inferior ones. Yet he was less extreme than many Victorian scholars, and he lamented the fate of those natives he saw as noble, such as the Maori of New Zealand. And his theory of evolution by natural selection was not only ahead of its time, but far ahead of the data that could test it properly.

Darwin saw evolution as a dynamic, visible process—even one that could be controlled, being responsible for new breeds of dog, racing pigeon, and domestic farm animal. He had no way of knowing how gradual the process of human evolution had been, or how many million years it extended back, and so it was natural for him to view human breeding like animal breeding, with immediately tangible results for good or ill. Because he had argued that our intelligence was evolved, and evolving, rather than God-given and static, he was sensitive to the fact not only that might it vary between different groups of humans, but that such variation should be expected.

Darwin wrote without demur that, around Cape Horn, "the different tribes then at war were cannibals."[34] They ate each other aggressively, as part of war, but also within a tribe, selecting the weak, the defenseless, and the no longer economically productive to be food when resources got critically scarce. Jemmy Button told Darwin that the victims were often old women, and they were eaten in preference to the dogs who could be deployed in otter hunting.[35] Darwin found Button's account independently corroborated by a Fuegian boy employed (presumably as a slave) by a local sealer, one Mr. Low: "This boy described the manner in which they are killed as being held over smoke and thus choked." Darwin considers that "Horrid as such a death by the hands of their friends and relatives must be, the fears of the old women, when hunger begins to press, are more painful to think of; we were told that they then often run away into the mountains, but that they are pursued by the men and brought back to the slaughter house at their own firesides!"[36]

Despite Darwin's account, and the fact that he was there, on the spot, able to observe and cross-question his informants, one of his modern editors, James Secord, tells us that "unequivocal" anthropological literature denies the possibility of cannibalism in Tierra del Fuego.[37] Secord cites no authority for his view, but it follows a recent fashion in scholarship that is repeated by Darwin's principal modern biographers, Adrian Desmond and James Moore. They write of "a gullible Darwin" who believed the Fuegian "joke" about eating relatives during famines (although they are careful that their picture of Darwin does not render him typically gullible, or there might be little value in any of the conclusions he reached with the help of native informants).[38] The problem here does not just have to do with a pick-and-mix approach to Darwin. Nor is it simply the inverted imperialism of quite a lot of modern sociocultural anthropology (where native peoples are allowed to be exotic only on the surface, while underneath they aspire to the same moral—and dietary, for it often boils down to the same thing—standards as people in the developed, urban middle-class worlds). More profoundly, it has to do with something that I have called "visceral insulation."[39]

Cooking adds energy to food, and the comforts of modern civilization reduce the amount of food we biologically need to burn metabolically in order to keep our bodies warm. But the technology that insulates us from cold and exhaustion also insulates us from the psychic rawness of nature. Visceral insulation is the trend toward disengagement from the actuality of hunting, killing, and gutting, or, in the case of domesticated animals, of rearing for the table and then physically dispatching—something which, in parts of the modern world, now requires an official license. We are no longer empowered to kill. And when we open a clean, white Styrofoam meal pack, our food—burgers, chicken nuggets, fish bites, hotdogs—no longer looks as if it has been killed. Many of us have never seen a food animal killed, and must try to imagine what it would be like. But we are unlikely to try very hard. And if we cannot easily imagine a pig squealing its last as it is dragged to the shambles to be converted into bacon, ham, pork, and blood sausage (or, more euphemistically, "black pudding"), how can we be expected to imagine that some tribal groups might have done the same, disarticulating articulate fellow humans?

Visceral insulation is a reverb of increasing technology and although it looks modern, we know that it has been around for over 5,000 years because from late-fourth-millennium Mesopotamian sites such as Tel Brak come thousands of ugly and cheaply made[40] artifacts known as "bevel-rimmed bowls" or BRBs.[41] These Uruk-culture artifacts are the earliest prototype of the Styrofoam fast-food container: sun-dried, thick-walled, conical-sided pottery bowls for holding food. Hardly pottery really, as they were pressed out of molds and left to bake hard in the sun. Hardly food really, as each was then filled, probably once only, with a basic, no-frills worker's ration of barley porridge. Sufficiently refueled to continue with the corvé labor that was needed to build large civic works on the Tigris-Euphrates floodplain—ziggurat temples and extensive irrigation systems—the worker (probably a slave) simply threw away the BRB (stacks of them have been found upside-down in the foundation trenches of large public buildings). The history of the first mass-produced food is interesting simply in itself. Yet its implications for a whole series of biological and cultural developments, belonging to both System 2 and System 3, are profound.

What the BRBs mean is that we know that there were by this time people in the world who had lost the direct chain of contact with the source of food that they ate. Placed in a relation of dependency with a complex institution, these workers of Uruk were physically reliant on the—by now highly complex—forms of entailment that supplied their city-state with raw materials and reproduced the essentials of human life. In this situation they were now exposed to other people's germs. Someone in the food procurement and preparation chain not washing hands (and such a practice may have been high on no one's agenda) would lead to outbreaks of gastroenteric illness on a scale never seen before—a scale that would leave an actual adaptive, genetic trace.

Just such a trace has been discovered by one of my former students, Ian Barnes, by tracing the frequency of disease resistance in urban populations in various parts of the world. At first glance, the SLC11A1 gene (formerly called Natural Resistance-Associated Macrophage Protein 1) has little to do with archaeology. But it demonstrates clearly one side of the artificial ape phenomenon through one of its variants, a so-called polymorphism known as the 1729+55del14 variant, which is known to provide a significant level of immunity to tuberculosis in humans. Barnes and his colleagues measured the frequency of this gene among the populations of seventeen cities. The cities were chosen to reflect different regional prehistories to test the assumption that where urban life had been longest established, disease resistance would be highest. The results were unambiguous: in Turkey, where urban-style living goes back 8,000 years, and in Iran, where it is 5,000 years old, the gene variant is far commoner than in Sudan, Lapland, or England.[42]

Tuberculosis was spread from cattle to people when cattle were domesticated and we started to live close to them; but it intensifies as people themselves start to herd together and as poorer classes begin to emerge, with impaired health. It would not have been spread specifically by bevel-rimmed bowls, but they were a key part of the unfolding interface between our cultural artifacts and our evolutionary biology.

The psychological effect of BRBs should not be underplayed either; now that food was prepared centrally, animals were killed centrally. Or, in fact, away from the center, as that is what a "shambles" was—the zone of the city where slaughter and butchery took place, out of sight. The appearance of this type of mass-produced fast-food vessel was part of an intensifying retreat from the wild, and an ever greater control over the terms of death.

The advent of farming, which had been the pre-condition for most urban civilization, had already ensured that cannibalism, which has a widespread signature in earlier prehistory, was already a far less frequent occurrence. For our Ice Age ancestors, it had operated either on a waste-not-want-not principle, where those approaching death would insist that their grandchildren honor their death rites with a good feed, renewing their strength for the next mammoth hunt; or, with equivalent but opposite logic, it was to do with brutal competition and aggression, denying your enemies the ability to support themselves and reproduce.[43] Farming began to make such behavior anathema: for the first time the dead were honored by regular burial in familiar-looking cemeteries, which were both a territorial claim to plow land and pasture and also a form of conspicuous consumption, showing that you had your meat supply thoroughly sorted without recourse to eating the dead.

And so, by degrees, visceral insulation within urban environments took hold, to the extent that in the developed world of the 1970s and '80s even the idea that people had once had customs involving regular consumption of each other seemed impossible, and the academic fashion emerged of seeking to "protect" the reputation of maligned tribespeople by deriding, disbelieving, ignoring, or minimizing any historical accounts.[44] But, far from the cities, rarely, and in the remotest places on earth, cannibal customs hung on into modern times. The academic naysayers have been peculiarly silent since November 13, 2003. On that day Ratu Filimoni Nawawabalavu, the chief of Nabutautau on the Fijian island of Viti Levu, made gifts of elaborately woven mats, cows, and carved sperm whale teeth to the family of the missionary Thomas Baker, whom his great-grandfather had ordered to be eaten, along with eight Christianized followers, in 1867.[45]

This apology—the third attempt at one—was accepted by Baker's descendants. Earlier efforts had involved the return of his boots, which the cannibals had tried, unsuccessfully, to eat.[46] But they had eaten all his other clothing, because it was seen as a part of the missionary. The act of cannibalism was to ingest a person, and a person is not a naked ape, but an artificial one. This observation perhaps tells us the absolute fundamental fact about aggressive cannibalism, informing its logic, psychology and symbolism. The person who is eaten is not really human, in the sense of being biologically part of a potentially inter-breeding species. As the enemy, they are, by definition, different. We may refuse to call it species difference, but it is at least specific difference. The person you kill and eat is not part of a community you voluntarily inter-breed with; the person is not, in fact, socially speaking, a person. They are alien. And their artifactual elaboration through tools and clothing, hairstyle and cosmetics, even the shape of the gifts they may try to give, are not seen as separate from what we would consider to be the unifying, underlying, naked, biology.

Looked at this way, the tendency to aggression in deep prehistory can be appreciated. Upright-walking hominins, with ever larger and more protein-hungry brains, vied for limited resources in the same East African environment. Biologically, we know that there were several species in competition: different tribes, in the biological sense. But after the beginning of technology, there were also different cultural traditions. Different groups made different tools. These, when dropped, lost, and recovered by the enemy, became alien technology. Like the tracks of specific prey animals, the shapes of stone tools had become potent markers of difference: different tribes, in the cultural sense.

CHAPTER 5

THE SMART BIPED PARADOX

Many theorists have viewed females as passive recipients of evolutionary change, relegated to the bearing, nursing and transporting of young.

—Lori Hager, *Sex and Gender in Paleoanthropology*[1]

THE STUDY OF HUMAN ORIGINS involves two interrelated biomechanical puzzles, which, in turn, resolve down into implausible energy equations. The first puzzle is why and how some apes first took to walking upright on two legs (bipedalism). The second is how some apes, having become bipedal, got super-large brains. The smart biped paradox, at its simplest, is that once apes started to walk on two legs, the pelvic girdle was constrained, so that the last thing we would expect, in terms of natural selection, is an increase in head size.

The promise of bipedalism is immense. Once a creature is standing upright, the hands are free to manipulate objects, and the position of the lungs and diaphragm in the torso is altered so that the breath control needed for complex vocal communication (speech) is potentially enabled. At the same time, the abdomen is squashed, and the gut length is reduced so that the surface area over which digestion can take place is sharply curtailed. This presents a puzzle in relation to the evolution of dramatically larger brains because, in humans, those organs are extraordinarily energy-hungry.

Bipedalism probably first began 5 or 6 million years ago in East Africa, but it may have emerged earlier on the island of Sardinia, in *Oreopithecus bambolii*, a Miocene ape that lived between 9 and 7 million years ago in a relatively predator-free environment where the upright position might have provided considerable advantages for foraging plant resources from higher in the undergrowth.[2] In East Africa, the presence of many more predators, such as the saber-toothed *Dinofelis*, which hunted the little *Australopithecus afarensis*—the Lucy creature—meant that the advantages of walking on two legs had to be balanced against an ability to nest in trees. So it is not until a little more than 2 million years ago, with the emergence of genus *Homo*, that we see the appearance of what paleoanthropologists call "committed terrestrial bipeds."[3]

What the initial advantages of bipedalism actually were is disputed. Some scholars strongly support the idea that being "hands free" was advantage enough (as it may well have been for *Oreopithecus*); others believe that the ability to cool the head by lifting it away from the ground (in a drying and ever more treeless, savannah-like environment), and spotting predators from an elevated vantage point, would have been critical. There is also support for the "throwing hypothesis"—that an upright stance allows a much greater range and accuracy for throwing projectiles. William Calvin has consistently argued that early chipped stone hand axes were not hand axes at all.[4] In line with the assessment first made by John Frere for the 400,000-year-old examples from Hoxne ("evidently weapons of war"[5]), he sees them as offensive projectiles (though primarily used for hunting rather than interpersonal conflict).

Finally, many scholars support the concept that endurance running, an activity in which humans outperform many potential prey animals, is central to the explanation. But as endurance running could not have come about all at once (it is an adaptation based on some form of preexistent and more limited bipedalism), we have to conclude that the sort of walking and running modern humans do may be the result of a progressive evolution driven by a number of requirements at different points in the past.

What has been almost scandalously ignored in the debate about upright walking is its most obvious and lethal disadvantage. Perhaps because we are so sheltered by the ubiquitous technology of baby carriages and strollers, we fail to grasp the extraordinary vulnerability of young mothers trying to transport their babies in a brutally wild environment. This is the smart biped paradox at its most acute: it is clear that the larger the braincase of the infant relative to its body, the longer it will take to support itself, and human children go through a painfully slow quadrupedal stage prior to mastering the complex balancing act that is walking. Human babies cannot even crawl at birth because, to a degree not seen in other mammals, their brain is incompletely formed at that point.[6] The most viable solution is to carry the child on your hip, cradled in one arm.

Although in monkeys and apes the young are not born fully developed, they are rapidly able to clamber onto their mothers' backs, or firmly grip their pelts, and the mothers are able to efficiently grip their hairy young back. Sometimes, if a troop is disturbed or attacked, young may fall from the trees or get left behind on the ground, but the dangers are far less than those that must have been immediately faced by an upright-walking, possibly hairless ape with infant offspring in open country.

Humans confound expectations in another way too. We have already spotted an objection to Darwin's analogy between the peacock's tail and human brain expansion. When a peahen produces male offspring, she does not have to give birth to something with a massive tail. She just lays an egg, and the chick emerges, the male and female looking virtually identical. The males grow the tails later. Humans, on the other hand, give birth to live young, with the head passing down the pelvic canal. In chimpanzees that canal is quite broad, and the baby chimp's head quite small. In humans, the pelvis, by virtue of needing to support an upright frame on two legs, is much narrower. Becoming upright should have led to a total freeze on brain-size expansion. Natural selection, as we are about to recognize, would have acted against it, as birth became ever more dangerous. And, if large heads were a sign of danger, a mark of potentially troublesome birth, then females, on any biomechanical logic, should have preferentially chosen smaller-headed mates. From this angle, it is equally hard to imagine a preference developing in terms of sexual selection, as Darwin supposed in his peacock's tail comparison. The peacock's tail is costly in terms of nutrition and maintenance, inefficient for flight, and it visually attracts predators. *Because* of all these disadvantages, the shimmering fantail signals to a peahen that its possessor has overcome all adversity and is a super-fit prospective mate. (And despite recent controversy over exactly which part of the overall display attracts the female—gaudiness, the number of eye-spots, or the accompanying calls during the mating dance—the males with the brightest feathers are measurably fitter than their brethren, as judged on a standard measure of the level of parasitic infection they carry.[7])

A peahen eyeing up a male with a massive fantail is considering mating with a super-fit survivor; it will have no downside in terms of the size of egg she will have to lay. The situation for a bipedal hominin female was fundamentally different. As the risks of bearing the offspring of a large-brained mate became apparent, female protohumans would have concluded that small, not big, was sexy; even if they did not, those that gave birth to smaller-headed children would have had an enhanced chance of basic survival.[8]

Let us look in more detail at the mechanical problem of large heads.

It should not be controversial to claim that, having evolved to walk upright, subsequent cranial development should have stalled, remaining at, or even moving below, the brain-size level of other apes. For around 3 million years, that is what happened (see figure 5 in Chapter 3).

Eventually we began to experience an increase in intelligence unprecedented in nature. We know it was a sudden evolutionary acceleration because, due to the intensity of research, the record of hominin cranial expansion has become quite full and detailed. Skulls and jawbones survive in the ground comparatively well, and are also more easily recognized than other parts of fossil skeletons, such as pelvises. But it is the pelvis that we really need to know more about.

Since Don Johanson's seminal discovery of a relatively complete skeleton—Lucy, an almost certainly female *Australopithecus afarensis* individual with well-preserved pelvis—only a couple of near-complete pelvises have been recovered.[9] The first, that of an adolescent male *Homo ergaster*, dating to 1.8 million years ago (the Nariokotome boy), was narrow, and suggested fairly problematic childbirth and/or small-headed infants.[10] But this is confounded by the more recent discovery made by a team led by Sileshi Semaw of Indiana University in Bloomington at the Gona site in the Afar region of Ethiopia, which had already produced the world's earliest chipped stone tools. The female pelvis from Gona is much later than the first tools, by about one and a half million years. Attributed to the species *Homo erectus*, it might be lineal descendant of the Nariokotome type in Africa, but it is remarkably different.

Figure 8 shows the different pelvic morphology of ape and hominin species, with gorilla and chimp to the left, their long, wide birth canals uncompromised by upright walking. In the middle is Lucy, with a brain still in the chimp size-range, but walking upright and with the pelvic opening oriented from side to side. The inferred birth position, with the infant head coming out side-on, is also the most likely for *Homo erectus*, as judged on the basis of the Gona pelvis. To the far right is the modern human pelvis, with its varying cross-sectional shape through which the fetal head must rotate if it is to emerge.

We know that Darwin's solution to the puzzle of the massive human brain was sexual selection. Women, he believed, were attracted to intelligent men, just as peahens were attracted to showy peacock tail displays.[11] Darwin reasoned that increased intelligence must have conferred some purely natural advantage at the start of our evolution, at the point when we pulled away from our primate cousins. But he could not see that as accounting for all the difference. After all, a human brain is three to four times larger, however it is measured, than that of any other ape; in other words, it is far more advanced than would be required simply to outsmart our fellow primates. In the mid-nineteenth century there was no way to get a handle on the precise timescales involved, but Darwin guessed, correctly, that it all took a very long time, measurable in at least hundreds of thousands of years, and that there was space for more than one process to operate. Earlier in our

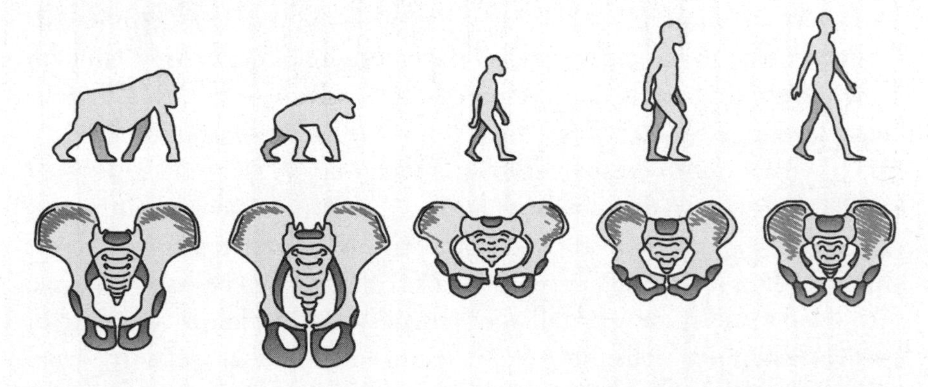

FIGURE 8 Pelvic morphology of ape and hominin species, from left to right: gorilla; chimpanzee; *Australopithecus afarensis* (Lucy); *Homo erectus* (Gona); anatomically modern human. (Graphic © Frankland.)

evolution, he concluded, it was the clever fellows who had devised spears and clubs and gone after the big game who were most reproductively successful. They gave their offspring an edge, and they, in turn, passed their enhanced capacities down to their children.

So, in Darwin's thinking, an association was established. Smart was sexy, and it brought the meat home. Long after brains had become big enough for basic hunting success, and what now turns out to have been several million years after we had out-competed our closest ape relatives in the battle to colonize the best and broadest ecological range, our brains just kept on expanding. If a man's hunting techniques got smarter and more subtle as his head swelled, that was all the more attractive to women. In addition, the female preference for clever mates would lead to intermale competition—another evolutionary arms race, but one based on the literal race to develop armaments. The man with the intelligence to devise the best extrasomatic devices—tools and weapons—would usually come out ahead. Such intense rivalry, in which the mighty hunter typically "gets the girl," would cause male brains to become ever larger.

Darwin did not consider that males might habitually want to choose the smartest females. Like the peacock's tail, which began as a flight stabilizer but stopped serving an aerodynamic purpose to the bird, the sexual selection process envisaged by Darwin for our human forefathers, but not foremothers, would have rapidly produced excess capacity over what was needed for basic survival—a runaway sexual selection whose by-products might include talents for art, dancing, singing, joking, playacting, mythmaking, and technological innovation.

On the basis of current archaeological knowledge, the first chipped stone artifact—the first artificial construct—came *before* the most significant stage in human biological development. Technology came on stage *before* we became large-brained. This tells us something that we have long resisted knowing. We did not evolve biological intelligence and then set out to create culture and technology. Darwin's guess that intelligence allowed prowess in hunting, that it made men attractive to women, and that it created competition among males might yet have truth in it (equally, the refined arts of campfire cookery may have attracted males to more intelligent females—but we know very little of sex roles in deep prehistory), but it seems unlikely that runaway sexual selection could have been the driver of the whole process.

One might have imagined that what would make least sense to Darwin was why any brain expansion had occurred *at all* once our ancestors started walking upright. For it is a simple fact of biomechanics that converting a quadruped chassis (the skeleton of the basic knuckle-walking ape, ancestral to chimps, gorillas, and us) into a fully hands-free biped meant that the pelvis had to become much narrower. The implication of this for any increase in head size in successive generations seems obvious: it should not have, could not have, happened (but it did). Natural selection means that the least risky births, the most live offspring, should logically be associated with the smallest heads. Darwin, probably because the actual evidence of our evolution spoke against it, did not spot the problem, and we also will leave it to one side at this point. But just because the huge increase in head size somehow happened does not mean that it is not an absolutely fundamental objection to natural selection as the explanation for the emergence of our species. We shall return to it after we have looked more closely at a puzzle Darwin did think needed answering.

Anyone who has cared for tiny human infants knows how much support they need. Not only can they not stand, but they cannot crawl and, if set upright, do not even have enough strength to support their own heads. Moving them any distance unaided is not an attractive proposition, and the early years in most of the world's cultures have their own specific-to-babies technology, such as the backpack we used when we took our first daughter with us climbing.

As we neared the summit of Green Gable, we sensed something wrong. Rebecca, whom I had been carrying, was too quiet in her backpack (see figure 9). I could not see her, but when I reached back and cupped the well-wrapped toes, the only movement seemed to be produced by me. She didn't kick back, and now Sarah came over and could get no response either. "She feels cold," my wife said. We scrambled fast now to lose altitude, dropping out of the bitter wind that had been buffeting us on the ridge, leaving the well-marked path with its thin scatter of late-afternoon trekkers behind. Off the path we could descend faster and gain shelter from the ridge above us. We reached a boulder field and squatted down out of the wind. The landscape around us felt suddenly vast, airy, and far from civilization. We were well enough prepared, with compass, map, and spare gear; but these were pre–cell phone, pre-GPS days.

As Sarah crouched in the lee of a large rock, cupping the head of our nine-month-old, offering milk, I realized that we had been dumped out of the high-tech modern world and were facing an archetypal challenge, the most basic one: keeping offspring as alive, healthy, and happy as their fully grown parents in a potentially harsh and sometimes unpredictable environment. This challenge is one that has often been downplayed in accounts of human evolution. Surely, the most critical thing that our upright-walking ancestors had to do as they moved over the savannah was manage to keep their infants with them. But how they did it, and what the parameters were, has until very recently remained so uncertain that it has been seen as unscholarly even to speculate.

As the paleoanthropologist Lori Hager has pointed out, this may be as much the result of latent sexism as of a real problem with the data.[12] Certainly, baby-carrying slings do not survive for archaeologists to discover in the same way as flint axes do, but neither do animal-fur cloaks, and everyone is happy enough to see those appear in colorful reconstruction pictures. Hager argued against the fondly held view that stone tools, invented by brave males for killing big animals and provisioning thankful families, were the earliest human technology. All the evidence points to humans needing to gather and carry plant foods well before stone tools were invented. "Females," Hager thought, "were likely to have been the earliest inventors as they developed tools in relation to gathering and in relation to carrying infants."[13] This idea, so clearly expressed in 1997 (in the introduction to Hager's agenda-setting book *Women in Human Evolution*), had been around for at least a decade previously, argued for by Nancy Tanner and Adrienne Zihlman primarily because they were rightly dismayed by an account of human origins that, ever since Mr. and Mrs. Darwin's backgammon sessions, had been "androcentric" (male-centered).[14] More positively, it was because they understood the implications of the primatological and ethnographic data on the way monkey and ape young were moved around and how (mainly) women in traditional hunter-gatherer societies did it.

What remained absent until recently was any clear understanding of the way in which human ancestors might have upgraded from a chimp mode of infant care and carrying to a human one. Did it happen by degrees or at a particular moment? If so, when? And what would the implications have been? It may not just be sexism that is to blame for the relative lack of progress. Perhaps it is also because we have all become so insulated from the realities of survival in the wild that we underestimate how critical it is to transport infants effectively and safely. There, in the mountains, with a tender nine-month-old, that abstract thought was a long way off.

Those who know the Lake District mountains only from the map might note their low elevation and imagine an easy stroll, not appreciating why it has been a nursery for some of the world's toughest endurance climbers. Green Gable presents a mere 2,628 feet of climb, ascent starting from near sea level, but that sea, which soon comes into view if cloud and mist lift for long enough, is the Irish Sea. Ruled by the North Atlantic system, at this latitude the weather can do almost anything, at any season, and it does, at a moment's notice. Every year, ordinary walkers die, many as a result of poor preparation and apparently undramatic falls. Unable to get off the hill, these unfortunates are equally unable to survive long on it. The light, the temperature, and the rain all fall without mercy. Exposure creeps up, sets in, and soon cannot be reversed. In 2006—a bad recent year, admittedly—eighteen walkers died (with twenty-eight fatalities in total, counting the more dedicated rock-climbers and others extreme sports fans).[15]

Every step forward is time that must be added to the safe margin for return before dark, and every small increase in altitude carries extra risk. At 3,209 feet, the highest Lakeland peak and the proud highest

FIGURE 9 Our first daughter, Rebecca, in a rucksack-style backpack. (Photo courtesy the author.)

FIGURE 10 Our second daughter, Josephine, in a classic cloth hip sling. (Photo: Sarah Wright)

mountain in England, Scafell Pike, seems only a little higher than Green Gable. But the extra 580 feet means I would not for a moment have considered attempting an ascent with our small daughter. I was left wondering how bad this misjudgment was anyway, as we were forced back within sight of our less ambitious objective. In the company of young adults, keen for the hills, I had learned about certain risks involving ill-fitting crampons, hangovers, incompetent map reading, the apparently trivial but potentially lethal issues of soaked boots, loose fasteners, and so on. But I was not experienced in this environment as a father.

I was climbing with my wife and our first child; the ascent up the steep Jacob's Ladder path from Langdale had taken longer than estimated, and I had begun to understand that, contrary to expectation, a live weight could be rather harder to climb with than a dead weight. The reverse is commonly held to be the case: the phrase "a dead weight" suggests that it is harder to carry. But that is because an active child (or adult), when being carried over the shoulders or perched on the hips, can adjust their center of gravity and aid the total weight distribution of the carrier in a way that a bag of coal or a sack of potatoes cannot. Yet, for very small children in backpacks, the reverse is possible. They can adjust in the contrary direction, and you never quite know what the effect will be. An unexpected movement as you reach for a handhold can send you off balance, and the carrying adult has to stay muscularly braced all the time.

Rebecca had been asleep, and her lack of movement had speeded our ascent onto the high ridge where the weather had changed for the worse. After her feed we strapped her back into the backpack and headed downhill as fast as possible. The dark ridges rose behind us, providing shelter from the wind even as evening entered the sky. Reaching a lower pasture, I felt movement behind my head, heard a rustle and a quiet voice: "Ta! ta!" Just after Becca's birth, I had brought a little gray woolen sheep toy back from Poland for her. It evoked the high Carpathians, and I named it Tatra. Now she was using the name to tell us about the sheep grazing the slopes to our left. Our adventure had produced her first words.

Cold shock sounds dramatic, but happens frequently in environments beyond the insulated protection of centrally heated homes. The onset of hypothermia requires only a two-degree-Celsius drop in core body temperature. Often dismissed as a bit of tiredness, a slight loss of appetite, it is slipped into all too easily and often not spotted. In children, because of their smaller mass to body volume, the drop in core temperature happens much more quickly (although, paradoxically, it may mean the condition is more survivable, as the rapidly cooled organs need less oxygen to stay alive).[16] We had discussed potential risks beforehand, aware that, despite warm clothing, inactivity could be an issue. That was precisely why we monitored our child closely and turned back with immediate decision at the very first sign that conditions were less than ideal. We knew in advance that we had the luxury of that choice; we were not compelled to continue into danger, like hunter-gatherers striving against time to complete a critical mountain pass ahead of snow.

In later years, and with our second child, we graduated to other forms of sling carrying, going, eventually, more low tech around town and in the hills. The Victorian elite, in harmony with the general disdain they displayed for interacting personally with their children, provided their wet nurses and nannies with perambulators (literally "walkers"). Soon these vehicles became a snobby "must-have" filtering down the social grades. Eventually even working-class women stopped carrying their babies in shawls and purchased baby carriages and eventually strollers.

Inheritors of the hippie resurrection of low-tech solutions, we were the first generation that began to be supported by commercially available alternative products. We were well-versed in what we thought of as back-to-nature parenting, but untrained ourselves in the best alternative technology. The around-town, front-facing sling that Rebecca used to travel in was wonderful, but complex. When Josie came along four years later, the range of options had expanded, and a simpler solution was at hand. We went for a variation on a standard African hip sling—essentially a wide band of strong cotton with marine-ring fastening that spared us from having to learn the correct knots (see figure 10). In the mountains I took to carrying Josie directly on my shoulders, holding either hands or legs: it had some new dangers, in that it set her higher, and gave me just one free hand (at most), yet it engaged her own sense of balance better (but beyond about forty pounds she had to do her own trekking).

The need to transport small children is central to humans. Richard Wrangham, in his study of energetics and the centrality of cooking, describes a group of hunter-gatherer Hazda women picking up their digging sticks to go foraging: "Some take their babies in slings, and one or more carries a smoldering log with which to start a fire if needed."[17] These essentials are identical to those of Aboriginal Tasmanians in similar circumstances; in their case the baby sling was made of the pelt of the Forester kangaroo, inviting the suggestion that in Australia the sling could have been copied from such mammals. Certainly, the sling turns humans, by artifice, into marsupials.

An essential element in the development of technology is the provision of substitutes for what exists in the animal world. A sling needs tying, perhaps sewing, and we should presume that, before bone needles were made, sharp bird beaks or the talons of birds of prey were liberated from their natural owners for human use, to puncture leather and thread through holes.

I forgot what was in my hand as I stepped down the shelving beach onto the perfect level of Silver Playa. Ahead, the peaks of the Avawatz Mountains rose out of oceanic brilliance, floating in a shimmering archipelago. As I walked forward, the water's edge hovered always a hundred yards ahead. Where this tempting mirage now lay fish once shoaled and birds skimmed the glistening waves. I tightened my grip on the sharp fragment of stone arrowhead. Many things can be achieved as expediently as I killed my fish in Oregon. Here in the Mojave Desert, on the edge of what, 10,000 or more years ago, had been a real lake where the Great Basin tribes came to hunt, I felt the edges of a small, beautifully knapped stone point. I wondered where the rest of the culture had gone to. The heat beat up from the dry salt lake, and I turned toward the beach, already much farther away than expected, and imagined myself wading through pellucid shallows, hauling a glistening catch back to shore. I placed the stone tip back on the gentle gravel incline.[18]

Place names tell you a lot out here—Silver Lake, Good Springs, Sandy, Lake Arrowhead, Big Bear City, Yucca Valley, Joshua Tree. Up to the north, beyond the Devil's Playground, lie Soda Lake and Goldfield. From the top of the beach the tatty charter bus came into view and parked up on the dirt road, and I rejoined my fellow archaeologists for our return to Vegas. While the others went back to the bus to rehydrate, I walked uphill until I was among the giant yucca plants, the Joshua trees that are endemic to the Mojave. Our guide had explained how to do what I was about to do, and also that it was not approved here in the National Park. But, like a typical human, I wanted to. The giant yucca has a woody stem and a radiating crown of succulent cactus-like leaves, each ending in a big sharp thorn. Furtively, I broke one of these thorns half away and, peeling carefully, managed to remove it with a long length of fiber from the edge of the leaf still attached to the plant. There it was: a Paleo-Indian needle and thread, sharp enough and tough enough to sew leather clothing. But archaeologically invisible.

In 1989, in her scholarly and readable work *Women in Prehistory*, Margaret Ehrenberg argued that

> one of the most significant human tools must be the container. Whether it be a skin bag, a basket, a wooden bowl or pottery jar, it allows us to carry items around or store them safely in one place. The container may have been one of the earliest tools to be invented, though unfortunately there is little archaeological evidence to demonstrate this. Chimpanzees can carry things in the skinfold in their groin, but when hominids became bipedal this skin was stretched and the fold was lost. The use of a large leaf or an animal skin, carried over one arm or the developing shoulder, or tied to the waist, might have replicated this lost natural carrier. One of the most important things that a female hominid would need to carry would be her young offspring. The complex interaction of bipedalism, food-gathering, the loss of hair for the infant to cling to, and changes in the structure of the toes which made them useless for clinging to its

> mother would have made it necessary for the mother to carry the child. The development of a sling for supporting the infant, found in almost all modern societies, including foraging groups, is likely to have been among the earliest applications of the container.[19]

Three years later Kathleen M. Bolen, examining the "prehistoric construction of mothering," reached similar conclusions: "women carry their infants (burdening), keeping them in close contact. An innovation, such as a strap or sling to facilitate transport of the infant, would not likely preserve, and such mention appears infrequently in archaeological literature. Baskets made to hold or encage children likewise would not leave archaeological traces. By overlooking or ignoring the potentials of such less archaeologically visible evidence, the infant always on the mother's hip continues to burden prehistoric women."[20] And in her wonderful recent account of mothers, infants, and the origin of language, *Finding Our Tongues*, Dean Falk, professor of anthropology at Florida State University, assumes that "A *Homo erectus* mother likely carried her helpless baby while collecting food, water, and other resources by using a baby sling," but immediately adds that "Unfortunately, we do not know when baby slings first appeared."[21]

All these scholars identify a critical issue for human evolution but come up against the problem I experienced in the Mojave Desert with the stone point and the Joshua tree needle and thread: we *know* that people, and the ancestors of people, used natural, organic materials most of the time in combination with stone tools, and the latter may have been only a small percentage of any artificial ape's day-to-day gear. Yet that is typically all that survives. At exceptional archaeological sites, like the prehistoric salt mine at Hallstatt in Austria where the Bronze Age wooden staircase was recently discovered, organic preservation allows us to look in detail at rucksack construction (see figure 11); the sophistication should not come as a surprise—Ötzi the Ice Man also had a pretty well-made bentwood and leather rucksack a couple of thousand years earlier. Such complex multi-material artifacts must have been long pre-dated by simple and elegant solutions for more general use, as represented by the ethnographic example of a whole goatskin with braided shoulder straps, from Mali (see figure 12).

But there may be another way of ascertaining when baby-carrying slings came into use among our ancestors, because, after we became committed terrestrial bipeds, that is, emerged as genus *Homo*, we were clearly adapted for endurance running. Even if, for some reason,

FIGURE 11 Original preserved Bronze Age carry sack from the prehistoric salt mine at Hallstatt, Austria. (Photo anwora © NHM Vienna.)

FIGURE 12 Contemporary whole-goatskin backpack from Mali. (Photo © H. Reschreiter.)

females with infants remained more static around the base camp, they would still have needed to move with the group. It should be possible to compare the available natural option of holding the child on one hip with the artificial option of using a simple sling in precise energy expenditure terms. Recent research by Cara Wall-Scheffler, Karen Steudel-Numbers, and coworkers has looked at both the changes in the length of the lower limbs in australopithecines and early genus *Homo*, and the calorific costs of infant carrying when upright.[22] Their conclusions are highly significant and show that natural carrying requires between 13 percent and 25 percent more energy expenditure than does using a sling, more than the average energy cost of lactation previously considered to be the major biological stress on young female humans and human ancestors. They believe that "the energetic drain of carrying an infant would be such that some sort of carrying device would have been required soon after the development of bipedalism and definitely to allow long distance travel, especially that out of Africa and across Asia."[23] So by 1.8 million years ago, when *Homo erectus* spread out of Africa, "tools must have been utilised to carry infants or other resources."[24]

Darwin was not the first to believe, on logical grounds, that humans must have evolved. The Greek philosopher Anaximander, born around 610 B.C., made the remarkable deduction that "in the beginning humans were born from other kinds of animals, since other animals quickly manage on their own, but humans alone require lengthy nursing. For this reason, in the beginning they would not have survived if this had been their original form."[25] Anaximander goes beyond a simple chicken-and-egg conundrum to draw attention to a unique feature of our species, the existence of childhood. It is possible that this pre-Socratic philosopher had examined early-term human fetuses. The Greeks were both intellectual and visceral and had scant regard for the lives of slaves or enemies. Certainly some such anatomical knowledge might account for his speculation that humans were originally "either fish or creatures very like fish; in these humans grew and were kept inside as embryos until puberty; then finally the fish-like creatures burst and men and women who were already able to nourish themselves stepped forth."[26]

This is an early, mythical version of a "phylogeny-ontogeny" argument. "Phylogeny" is the term now applied to the evolutionary development of species as they emerge and diverge. "Ontogeny" is the sequence of developmental stages that the fetus passes through between conception and the moment it is ready to be born. So the modern biological maxim "ontogeny recapitulates phylogeny" means simply that the stages of fetal development mimic those of the species' evolution.[27] The similarities in the initially legless embryos of both fish and mammals, and the way the buds that will become fins in the former become articulated legs in the latter as fetal development proceeds, do indeed reflect evolutionary relationships among vertebrates (even though the much-vaunted "gills" seen in human embryos are not actually gills but slits that in fish develop into gills and in humans into pharyngeal pouches).[28] Anaximander's fantastical account of human development sees things the other way round. Yet, his massive guesses are informed by a prescient focus on our oddity in not being self-supporting at birth—a feature that remains a crucial clue to comprehending how we actually evolved.

Moving from Anaximander to Mickey Mouse, we approach the first part of the solution to the smart biped paradox. "Neoteny" is a biological term for the retention of infantile characteristics into adulthood. The evolutionary biologist Stephen J. Gould, in his intriguing article "Mickey Mouse Meets Konrad Lorenz," argued that, through progressive redrawings since his "birth," Mickey had become ever more neotenic.[29] He went from a small-headed, dot-eyed, pointy-nosed, spindly, ratlike mouse in *Steamboat Willie* (1928) to the cute, chubby little fellow directing wizards' magic in *Fantasia* (1940). Mickey became cuter because he appeared more like an infant, with large eyes, domed forehead, and receding chin. These press all the hardwired evolutionary nurture buttons that Darwin discusses in his profoundly significant work *Expression of the Emotions in Man and Animals*, with its engravings of sulking chimpanzees, crying children, and happy puppies.[30]

Many animals besides cartoon mice have been claimed to retain infantile features into adulthood, but there are disputes around precisely how these features should be defined. What is clearer is that human babies in the first year of life are, as the Swiss zoologist Adolf Portmann argued, essentially extra-uterine fetuses, displaying what is known as paedomorphism.[31] In comparison to other apes, we are born prematurely, and this can be measured very clearly in neural growth, which, while it slows sharply after birth in chimpanzees, continues at a rate of a quarter of a million neurons per minute in human babies until the age of one (see figure 13).[32]

This fact has immense implications for the way in which human technological culture, whatever it happens to involve at any point in history or prehistory, can become essentially hardwired into the neuronal architecture—the essential biology—of our species. But before we look at the implications, we need to understand how humans evolved to a point where they could afford the luxury of producing offspring that (for all their wonderful potential) are so pathetically useless at the outset. We now turn to the second part of the solution to the smart biped paradox. Changes in the anatomy of our human ancestors, as discerned by paleoanthropologists from the preserved fossil remains, correlate poorly, if at all, with what the archaeological record tells us. As Ian Tattersall neatly puts it, "while the patterns are similar the two records are distinctly out of synch: biological and technological change did not proceed hand-in-hand."[33] There is a lead and lag in these things, and it is my contention that the precondition for the development of human intelligence was the presence of a fractionally but significantly more

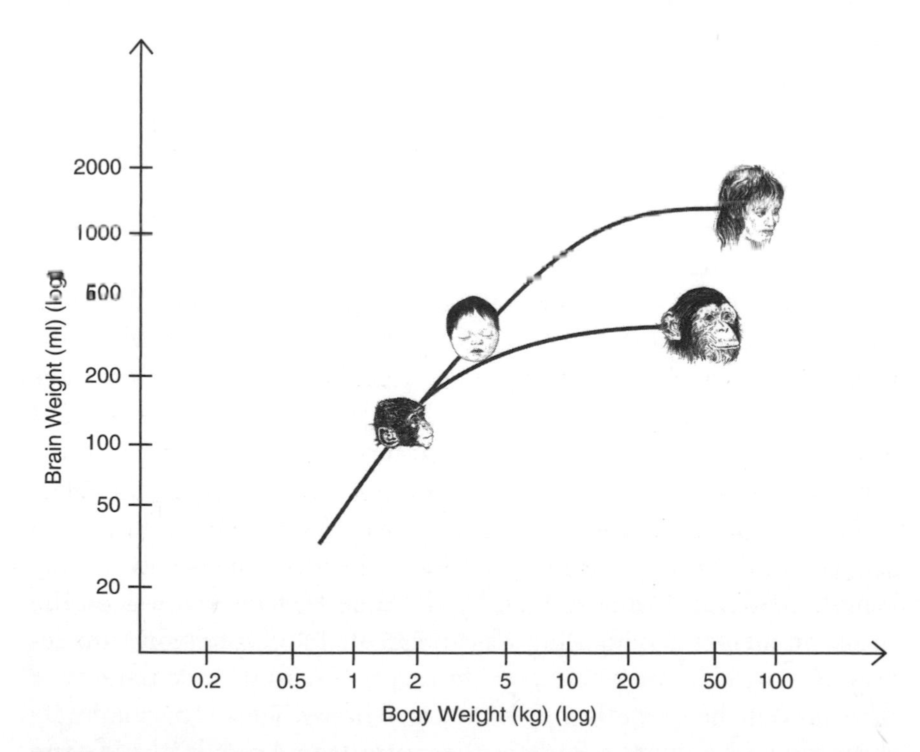

FIGURE 13 Comparison of brain growth rate before and after birth in humans and chimpanzees. (Graphic Frankland/Taylor; data from Bogin 1997.)

complex material culture among australopithecines than that observed by modern primatologists for gorillas and chimpanzees.

We know that by 1.5 million years ago, children with heads in the modern size range could, theoretically, have been born through the modified pelvis of *Homo erectus* females like the one from Gona, but how we reached this point begs many crucial questions. It appears that, having managed to stand on two feet for at least some of the time 5 million years ago, we remained partially tree dwelling until the first technological culture emerged around 2.6 million years ago. Pressing survival problems needed solutions: how to transport young safely; how to protect groups, day and night, in an increasingly open savannah; how to get enough food suitable for a shortened gut. And solutions were found. Fire began to be used, first at source—accidentally cooked meat might be scavenged, or protective fire brands obtained; later fire itself was scavenged, to be kept going in discrete locations to which branches could be dragged. Stones began to be made systematically sharper, and thus more damaging and effective as projectiles. Small, handheld slings might have increased the range over which they could be fired, sacrificing accuracy for distance and amount. But if predators were not forced to flee in a hail of sharp stones, the group had to move on, with infant children riding in larger slings at their mothers' sides.

All that actually survives from this time are the stone tools, and the knowledge, from Taung, that large predatory threats existed. We should suspect that the artifacts we do have were regularized in shape for a reason, and that the reason was that they were a part of an incipient process of entailment. That is, the ever more careful production of tools, from Mode 1 to Mode 2 (see above, Chapter 3, and figure 7), although pre-dating the later varieties of stone artifact that we know had to be combined with organic materials to actually function, indicates that they were quickly becoming critical in the production of other things that have not survived archaeologically. Not just critical in such production, but part of ever more joined-up technological routines. A stone tool used to scrape clean the insides of an animal skin could, if the skin was held in a loop, then be launched as a lethal slingshot from it. We infer that, by the time *Homo erectus* was on the scene, an infant-carrying sling was an essential tool, and stone projectiles of some kind must have been in regular use to provide the sort of high-protein diets that the much-enlarged brains, and correspondingly shrunken guts, required. Perhaps the sparks from knapping had begun to be used at this point to kindle the cooking fire *ab initio*.

We can now define a series of stages, starting from the natural primate state when the mother uses all four limbs to move and the infant has to ride on her back or hang under her body. The infant has to gain competence in mobility quickly after birth, limiting the time spent statically nursing the newborn in some arboreal nest, if mother and child are not to be vulnerable to predators. In chimpanzees, the infant has to be carried by and cling on to its mother for around the first five months,[34] and learned skills in mothering are already critically important (Jane Goodall noted that inexperienced chimp mothers lost more than 50 percent of their offspring, while in the most experienced it dropped to "just" 17 percent[35]).

The next stage, corresponding to the development of limited bipedalism, involves hip carrying, with the infant supported by the mother's arm or arms, which, thus encumbered, are hampered in other activities: for females at least, some of the "hands-free" benefits of bipedalism are undermined by the need to transport young and keep them safe. These two stages are biological only, and then comes the breakthrough: an inspired female picked up a twisted loop of animal skin—perhaps some sun-toughened membrane at a scavenging site, or some scorched but not burnt pelt from the embers of a bush fire. With the infant seated and the strain off the arm, energy requirements for moving with the child plunge, by a massive average 16 percent. So the pressure to make this discovery, even to remake it more than once, is huge. It becomes conceivable that the first bespoke and standardized stone tools in the archaeological record were made in order to obtain the materials for and complete the simple fabrication process for basic slings. Any group that could develop the basic technique to secure the growing child behind them, in the rucksack position—a standard orientation for longer journeys with infants in many parts of the developing world—would have made even greater energy savings. The emergence of the sling is a way of adding energy into the system in advance so as to save more of it later. Just as Richard Wrangham demonstrates with the added calorific value that comes from cooking prior to eating, so taking the time to make a sling pays dividends when walking the next day. In modern technological societies, large amounts of energy are pumped in ahead to manufacture metals and plastics, wheels and rain hoods, and all the components of a baby buggy or stroller. And, of course, we can also put infants in cars and on planes and, in hours, move with them distances that our *Homo erectus* ancestors took thousands of years to cover.

Although carrying slings do not in themselves drive brain-size increase, they certainly encourage it. Rather than having to fit a larger and larger cranium through a pelvic girdle that has contorted itself to support an upright frame, helpless babies can be catered for in a pouch. The time that they remain helpless—a week, a month, a year, several years—becomes less critical. So sling technology removed the glass ceiling on the degree of ontogenetic retardation (in primate terms, premature birth) that genus *Homo* could begin to accommodate. And that, of course, is the solution to growing larger brains: you do it outside the womb.

From the point of onset, our evolution was critically driven by technology. The sling, in particular, allowed the final australopithecines to become artificial marsupials and, by accommodating ever higher levels of helpless paedomorphism, took the lid off the pelvic limits to bipedal brain expansion. Increased intelligence, in turn, allowed the development of a more complex techno-cultural system, the evolution of a fully and habitually two-legged stance, and the possibility of colonizing all the continents of the world.

We all hate uncertainty. Even in Vegas you can see that—games of chance can be fun, but if you don't know where the limits are and haven't booked the ride back out of town, it is a miserable descent into the unknown. So even the roulette wheel is something we have to be certain about. We have to be sure the ball will land on a number by pure chance, and that is a special kind of certainty. When we pick up a book, flick a switch, press a button, or turn a key, we want the result to be predictable. And a roulette wheel is predictably unpredictable. Human evolution is like a game of roulette: unpredictable in a predictable way, and with two sorts of outcome: red and black.

Let the red numbers be the biological outcomes. These are the stages by which we moved from being tree-living, small-brained, furry apes to being naked, upright-walking, large-brained humans. It did not happen all at once, but stage by stage, result by result, not always in a predictable sequence. Let the black numbers stand for our cultural development, the things we have invented, from stone tools to houses, wheels to satellites, shoes to spectacles. These have been just as much a part of our story as the red numbers, and as in a roulette game, sometimes the red and black alternated and sometimes there was a run of red numbers. Right now there is such a run of black numbers—inventions that change our lives—that it can seem as if the red numbers have stopped. But biological evolution is still going on; the red numbers are still there. In the distant past, we like to think it was all red numbers, apes physically morphing into modern-looking humans. But that is where we underestimate the black numbers, the critical spins of the wheel that kept us in the game.

Between 8 million years ago and 2 million years ago we went from being tree-living apes with small brains to being savannah-living humans with large brains. We went from moving along the ground very infrequently to being on the ground all the time, and along with that, we evolved from some sort of basically quadrupedal or four-legged gait to a fully upright, two-legged or bipedal, stance, the posture characteristic of the hominins, of which *Homo* is the only surviving genus and modern humans are the sole remaining species. And there is a profound paradox right there.

Fundamentally, it has to do with the shape of the pelvis and the different job it had to do when we moved from being long-armed, short-legged tree-swingers (brachiators) to being habitual savannah-dwelling bipeds, with long running legs and relatively short arms. Up in the trees the pelvis only provided part of the support for the body. The arms took much of the strain through the shoulders, with the backbone hanging between the two, now supported at one end, now at the other, now at both ends: swinging, landing on a branch, steadying, relaxing into an all-fours position, leaping up and swinging again. Down on the ground, everything is changed. To be habitually bipedal—to walk and run as the main way of getting around—means the backbone has to perch on top of the pelvis.

To demonstrate the smart biped paradox, imagine a deck of cards, some roulette chips, a roulette ball, and a steady hand. A couple of card holders will be useful too, and a golf ball and scissors. We start by stacking roulette chips. Under my rules for the classic challenge, four large, yellow, high-denomination chips are placed first, followed by five purple, five pink, twelve green, five white, and, lastly, two white with gray insets. The yellow ones and the purple ones can be glued together to make two solid blocks: these then represent the vertebrae of what used to be a tail (the coccyx or tailbone in humans) and the fused vertebrae of the sacrum that completes the basket shape of the pelvis from behind; above that, the lumbar, thoracic, and cervical vertebrae have to be balanced one by one, culminating in the two white-with-gray inset chips representing the final, specialized neck bones, the axis and atlas vertebrae, upon which the head sits. Now we have to make the first stage of a house of cards: two uprights (let's put them in card holders to make it easier) and one across—the ace of hearts—and balance the column of chips on top. If the card holders are placed too far apart the flat card will bend and the stack will fall. If the flat ace of hearts stands for the pelvis, then I can cut out the heart in the middle to allow a roulette ball (the fetal head) to pass through. It is clear now that the two upright cards can be positioned just to either side of the aperture, and the stack of chips (the vertebral column) shifted backward a little to expose the central opening.

This is not yet Las Vegas *Homo;* this is only Las Vegas *Australopithecus.* Practically speaking, what we have is a very good model for the earliest type of savannah-dwelling biped. The backbone is no longer supported from both ends as it was in our tree-living ancestors, and has to balance upright, subject to new strains. The pelvis, which in chimpanzees was a fairly capacious bony girdle off which the stocky, wide-set hind legs hung, has now been forced to narrow to form a platform. This platform allows the weight of the torso to be conducted through the backbone down the legs and onto the ground. The choice of the ace of hearts is neat because the only surviving near-complete australopithecine pelvis, that belonging to the female *A. afarensis* known as Lucy, does indeed have a small, heart-shaped canal.

As long as the backbone does not have to support a heavy head, this will just about work. The head will have to remain small because the legs have to meet the pelvis as closely as possible directly beneath the backbone. In Las Vegas *Australopithecus* this configuration centers on the little, central ace-of-hearts-shaped hole around which the upright elements, both above and below, have to be positioned. But for Las Vegas *Homo,* instead of a roulette ball we need a golf ball. Obviously, it will not fit through the ace of hearts. So I replace it with the ace of spades. This card stands for the early human pelvis. With the big black splotch of ink in the center cut out, it remains questionable that the golf ball will fit through. It may be a push, but then that is how human childbirth is—far harder than it is in living chimpanzees or (as far as we can tell) extinct australopithecines.

The practical problems for the Las Vegas simulation are now critical: Will the stack of chips stand up in the first place? I need to move the base of the chip pile back as far as possible, and maybe I will also need to stack them in a slight inward curve, corresponding to the actual curvature of the human spine. The card holders, the feet, will be forced much farther apart, and our ace, the pelvic girdle, weakened by the cut-out, will start to sag and bend. What was a practical, if rather eccentric, simulation model, my Las Vegas *Australopithecus*, has become a massive engineering challenge—the bizarre conundrum of Las Vegas *Homo*.

Because humans exist, it is hard to see why we should not. But the smart biped paradox is exactly this: having developed upright walking, with all the consequent biomechanical demands on the pelvis, the *very last thing*, evolutionarily speaking, that should have happened was that this creature should have begun to expand its head size. In fact, for well over 3 million years, logic prevailed. The bipeds managed an increasingly open existence, with less and less tree swinging and more and more running about on two legs with nothing more than a static or shrunken cranial capacity. Enhanced intelligence should never, ever, have emerged from a two-legged primate. And then it did.

Between 2 and 1.8 million years ago there was a rapid encephalization: brain size, relative to body size, suddenly began to shift, not downward to facilitate easier birth but in the opposite direction, the impossible one—upward. And once it began to shift, there seemed to be no stopping it.[36]

For a very long time, evolutionary biologists have looked at this shift and seen it as a red result on the roulette wheel—a biological change, caused by genetic changes and positive selection for a larger brain. What they did not see was how impossible it was (after all, it happened). At least, impossible on a wheel with only red numbers. But the human roulette wheel has black numbers too. Around half a million years before the brain-size increase there is a black number: the first stone tool. Could it be that, somehow, this black number result altered the probabilities, allowed something to happen on the red numbers that had previously been precluded?

As I said at the start, I want this argument to be predictable. Throughout this book we see black numbers followed by red numbers and red followed by black. What I argue is that the usual way we have viewed human evolution, from the very earliest times right up to the present, is wrong. The usual view is that the red numbers always come first (the biological capacities of increased intelligence, better eyesight, more sensitive hands, and so on) and the black numbers follow (stone tools, the invention of fire, the discovery of agriculture). I want to show that, as with stone tools and bigger brains, the tools came first, black before red. In fact, black enabled red to happen.

Our genus, *Homo*, has a long prehistory, and is everywhere associated with the production of technology. The longest sequence at a European archaeological site dates back 1.2 million years, at Atapuerca in Spain,[37] but in Africa technology appears to even pre-date humans. The oldest so far identified stone tools date to 2.6 million years ago, when we know that *Australopithecus africanus* is beginning to show an increase in intelligence above chimpanzee level; but, as the talon-punctured brain case of the Taung child demonstrates, it was still struggling up the pecking order in a hostile environment.[38] But that was in southern Africa, where this species may have formed an enclave. In Ethiopia by this time, a very similar creature, *Australopithecus garhi*, shows up on the sites with the earliest chipped-stone tools and earliest systematic meat processing. The legs of horses were jointed using stone tools, presumably to allow them to be roasted easily, while antelopes had their tongues cut out and their legs smashed to get at the nutritious marrow in ways that are familiar from hunting butchery among hunters in the modern world. Surely carcass-processing as systematic as this involved an equally careful and organized use of the products, not just meat, but sinew, hide, and fur.

In *The Dawn of Human Culture*, Richard Klein and Blake Edgar say

> if stone flaking and brain expansion were closely linked, then brain expansion must have begun by 2.5 million years ago. Future discoveries may confirm this—or they may not. The *Australopithecus garhi* skull from Bouri, Ethiopia...provides fodder for doubters. This is because it anticipates *Homo* in its dentition, but not in the enclosure for the brain, which was no larger than in other australopiths. The Bouri deposits have not yielded any stone artifacts, but they have provided animal bones that were cut and broken with stone tools. Unlike nearby Gona, Bouri lacked cobbles or other rock fragments that were suitable for flaking, and when the tool makers visited, they may have carefully conserved their implements until they could return to a locality like Gona.[39]

Although researchers like Tim White are holding out for the possibility of a larger-brained creature being responsible,[40] an early member of genus *Homo* that has not turned up yet, the balance of evidence is on the side of the technology preceding the revolutionary biological changes that will lead, eventually, to us. Obviously, all sorts of evidence could still be out there. It almost certainly is. But what I want to suggest is that there is nothing inherently back to front in seeing things this way. In fact, even if a larger-brained creature did turn up, I would still maintain the need to have technologies preceding it, aiding those hominin females who had at once to compete with a wide range of competitors, including other hominins, in open country, while accommodating large fetal heads and ever-increasing neonatal helplessness.

CHAPTER 6

INWARD MACHINERY

It is the Age of Machinery, in every outward and inward sense of that word.

—Thomas Carlyle, *Signs of the Times* (1838)[1]

THE DISTINCTIVE RED FASCIA of the refrigerated Coke machine was unusual, unique perhaps, with its iconic Ice Age hunting imagery. It would turn out to be the closest my family and I would get to the herds of aurochs that roamed Altamira in prehistory, wild cattle that were both eaten and captured visually by the people of the Paleolithic. Around 15,000 years ago, artists working with soot in flickering torchlight deep underground had created the now world-famous "chamber of the bulls." We knew in advance that the cave at Altamira had long entered an enduring semiquarantine to halt damage and stabilize conditions after many years of exposure to raised light levels and the moist, bacteria-laden breath of twentieth-century tour parties. But we did not know that the replica cave in the air-conditioned unit adjacent to the visitor center was almost as inaccessible, being booked for days ahead. We watched the buses arrive, disgorging the passengers who compliantly donned headsets to await enlightenment in a variety of languages. As a prehistorian, I might someday have a chance to be included in one of the very few, highly regulated groups who still enter the actual cave. But I have no plans to revisit.

Each *tour de force* subterranean bull image was completed without room for error or correction, by a virtuoso cave painter. Each Coke can was mass-produced on an assembly line, filled by machine, and may not have been touched by human hands until opened. The distance between the two seems immense, yet the Paleolithic mammoth hunters also mass-produced near-identical artifacts. Their stone tools were made individually, but in strict conformity to standard models. They would have understood the idea of the Coke can.

I did not actually know what went on inside the Coke machine when I placed the correct money in the slot, but I trust to a learned causality. The Paleolithic cave artists did not actually know how their sacred paintings were connected to the actual wild aurochs, whose seasonal movements and annual calving they depended on. It would be easy to say that their paintings and their aurochs hunting were not causally related, but they almost certainly believed that they were. And in many ways, they had a stronger causal understanding of the interface between the natural and the artificial than I have. They made all their own technology, from spears to baby slings, pigments to shelters, beads to fall-traps. I would not have much idea of how to make a Coke machine from scratch, or the alloy cans inside it, or the coins needed to free them. My machines are perhaps as mystical as was their nature.

The cookie cutters had different shapes: a rabbit, a heart, a long-necked dinosaur, a star, a shooting star. Becca, my elder daughter, then eight years old, was unusually silent as she arranged them in a row on the windowsill. Satisfied, she announced what it meant. The shooting star was "the big bang"; the star was the stars; then came dinosaurs; the heart was "humans mating so there are more of them"; finally, there were pets.

This creation story, made with a few ready-to-hand symbols, had an impressively compact accuracy: the beginning was the beginning of time, followed by the consolidation of stellar bodies, the emergence of humankind (with the romantic symbol of the heart understood in unabashed reproductive terms), and all this grandeur and potency led to the little pet rabbit. Cute and fluffy, it was an icon of domestication, a sign of human ascendancy over nature. All of us carry some sort of "just so" narrative in our heads, some set of images of how the world has come to be. Some are very scientific, some are overtly antiscientific, some mix faith with reason. I am wary of making a direct equation between the creative lives of children and those of our distant ancestors, but I could not, and still cannot, help sensing kinship between the lined-up cookie cutters and the reasons for the very first art. Both of them are about creation, about why things are.

Few subjects have been so heavily discussed as the animal paintings of the French and Spanish caves. They may not be the very first art that humans were capable of, but they are the earliest surviving system of signs, symbols, and icons through which some kind of story seems to have been told. They have survived for 15,000 or 20,000 or, in some cases, over 30,000 years owing to their remoteness, inaccessibility, and (not to be overlooked) the superb quality of the materials chosen and the techniques employed. Each generation of archaeologists has had its own particular explanation for the bison, the stags, the mammoth, the strange humanoid figures, the geometric zigzags, the mouth-sprayed rows of dots, the hand stencils with their missing fingers. It is sympathetic magic for the hunt; a place of initiation; a picture guide to the gendered construction of the world, of male and female; a dark, hallucinated world of drug-induced trance phenomena. Perhaps it is all of these.

Certainly, in the most specific way, we shall never know. It is no longer given to us to understand this art in its original context. We are not those people, although we can reconstruct their patterns of life with increasing detail through many decades of painstaking excavation, recording, and analysis. Perhaps the mammoth pictures are meant to depict the very first mammoth, the mother or father of all mammoths. Maybe it represents all mammoths, or one particular, real mammoth, which became attached to a particular human history in a similar way to Jumbo, the Sudanese elephant who in 1882 joined Barnum's circus via the Paris and London zoos and whose ashes are now kept in a peanut butter jar at Tufts University, which institution he now symbolizes, while also becoming a shorthand for elephantine dimension (Jumbo-size). Perhaps, like Dumbo, Disney's variation on the theme, the cave art mammoth had special powers—clairvoyance rather than flight.

Maybe—and this is highly likely—the mammoth that I saw, painted in red and surrounded by little signs, in the cave of El Pindal in northern Spain, meant something a little bit different (potentially a lot different) from the one I saw painted in black, on the wall of Peche Merle in France.[2] The mammoths in the cave of Chauvet (which I have not visited) are perhaps 32,000 years old, while that in El Pindal may be only 18,000 years old, and the question arises of continuity of meaning over such immense spans of time.[3] I am not going to suggest any very particularistic meaning for cave art, or a set of locally valid meanings, the stories of particular caves, and neither am I going to try to weave a way between the various interpretations that other archaeologists have put forward.

I am going to suggest an overarching explanation for the Upper Paleolithic art, into which most of the other interpretations could happily fit. Not vague, not wholly novel, hopefully persuasive. The art, no matter in which cave it was done, or when, is about creation. The clue is in itself, in the act of making it. The stories told around it, before and after the event of painting it, may have varied greatly. But the underlying sense is inherent, as obvious to us now as it was to them then, if we can open our eyes naively enough to really see.

The human symbolic revolution has been characterized by the prehistorian Colin Renfrew as "the sapient paradox."[4] How can it be, Renfrew wondered, that, although humans emerged biologically nearly 200,000 years ago, it is only in the last 40,000 years that they show signs of complex symbolic behavior? This is an immensely challenging question.

Ice Age art appears at a critical juncture in our relationship with technological objects. While many human societies continued with a more or less non-entailed Aboriginal Tasmanian ethos, some found themselves caught up in an increasingly complex network of objects and relationships that started to take on a life of its own.

One answer to the sapient paradox may concern the archaeological record. Like the Joshua tree needle, the stone fish killer, and the infant-carrying sling, the earliest symbolic culture may often not have left permanent traces. The recent discovery on an Iberian Neanderthal site of marine scallop shells with the remains of prepared pigments inside them strongly suggests that Neanderthals practiced the cosmetic arts, most likely body painting.[5] But they might also have painted pictures on exposed surfaces or decorated the organic components of their shelters in ways we can now no longer know about; we know that many human groups, such as the Aboriginal Tasmanians, although they did rock paintings, also practiced their symbolic culture in ways that were more or less transient and unlikely to survive long in the archaeological record.

Nevertheless, it is clear that the European Upper Paleolithic, from around 40,000 or 35,000 years ago, saw what is known as the "creative explosion," with highly artistic ceremonial burials, extensive carving of bone, ivory, and stone, and remarkably complex and skillful friezes of paintings in a set of caves in southern France and northern Spain.[6] Engravings on open surfaces are known from Portugal and in British caves, and engraved stone plaques have been found in the Rhine valley. Whatever meaning or meanings attach to this creative explosion, all interpreters agree that it relates to some form of ritual, religion, and/or cosmology. Central themes appear to be the difference between the human sexes, with a contrast set up between carvings of phalluses and those of vulvas, and between elaborate ceremonial burials of male bodies decorated with red ocher and miniaturized carvings of female bodies, some also coated in red ocher. At the same time, stone tool making was becoming dramatically more entailed. Supply networks for the best-quality raw materials now extended across Europe, and very little of the technology was any longer expedient.

French archaeologists studying the Paleolithic Ice Age cultures from the 1960s onward began to think about the processes involved in making gear for a given purpose by analyzing the *chaîne opératoire* (operational sequence) embodied in particular stone tools.[7] They looked at where the raw material came from (fine imported flint or local stuff good enough for most uses?); how the tool was made (in how many stages? chipped in a basic way or "retouched" in a more painstaking fashion?); how it was used and reused (kept a long time and gradually reworked, so a large axe ends up as a small scraper, or used a few times and thrown away?); and where it was used (butchering an animal where it had been killed, or back at base camp?).

One way of understanding the shift to more entailed technology has already been outlined, in the classification of forms in the Mode 1 to 5 sequence. Another way of looking at it is in terms of the relationship between mass of raw material and degree of function measured along a specific dimension, such as extent of usable cutting edge. The French Paleolithic researcher André Leroi-Gourhan calculated that one kilogram (about 2.2 pounds) of flint prepared in the earlier Paleolithic Olduwan (Mode 1) tradition could be roughly chipped to make a 10cm (four-inch) long blade edge. By the Middle Paleolithic, tools such as the ones John Frere found at Hoxne—Acheulean hand axes—might provide 40cm (fifteen inches) of cutting edge, with two hand axes produced from the single flint lump (these types actually cover Modes 2 and 3). By the beginning of the Upper Paleolithic and the symbolic revolution, the same amount of flint, reduced through the procedures of prepared-core blade technology (Mode 3), could produce two meters (six feet) of cutting edge, and by the end of the Ice Age, in the final hunter-gatherer period prior to farming, up to twenty meters (sixty-five feet) of edged bladelet (Mode 5) were possible.[8]

While the Acheulean hand axes were either standalone or entailed simply (for instance, for use with a sling), the later, Mousterian (Mode 2) and Magdalenian (Mode 3) blade technologies were frequently designed to produce the elements of multipart tools. Production had become highly complex at this stage, with long apprenticeships and perhaps some specialization in skilled activity, with complex chains of procurement, and the need to bring many materials together at one place: wood, bone, and antler for armatures; pitch, tar, and fish glues for fixing; feathers for fletching; sinew for binding; and so on. And that would be simply to make an advance projectile, like an arrow or stone-tipped spear, which would itself need other technology, such as an elaborately carved spear thrower or a yew-wood longbow, like Ötzi's, to make it functional.

As things became more and more entailed, with a complex operational procedure or *chaîne opératoire* involved in the fabrication of each element of each tool, we should not be surprised to see ritual and religion becoming visible in the shape of sacred art and elaborate disposals of human remains. This appearance does not mean that our ancestors had suddenly become smarter than they had been (of course, according to some modern commentators, developing a belief in supernatural entities might make them appear stupider). What it does mean is that two interrelated factors had come to the fore, rote learning and risk aversion.

Rote learning was imperative to the mastery of various *chaînes opératoires*, whether for stone tool production, the weaving of textiles (which are believed to have grown more complex at this time[9]), the preparation of pigments for body or cave painting, or the sculpting of detailed figurines. Rules for making things multiplied because further rules governed using the things you had made as tools to make other things, which might, in turn, produce further things. But a mistake anywhere along the line, a missed step in the sequence, and the end product would fail, perhaps at a critical moment. It is human nature to try to avert such failures.

In the modern world, approved standards and procedures are written down as blueprints, protocols, recipes, and equations. These can be analyzed logically after any failure to identify the reason. But that does not stop many of us from crossing our fingers, or engaging in a wide range of other little obsessive-compulsive spectrum activities to calm our nerves. Back in the Upper Paleolithic there was no writing,

FIGURE 14 Upper Palaeolithic figurines from the site of Willendorf II, Level 9, Austria: the famous limestone Venus of Willendorf is flanked by the less distinctly formed mammoth ivory Venuses 2 and 3 (24,900–23,900 years ago). (Photo: A. Schumacher © NHM Vienna.)

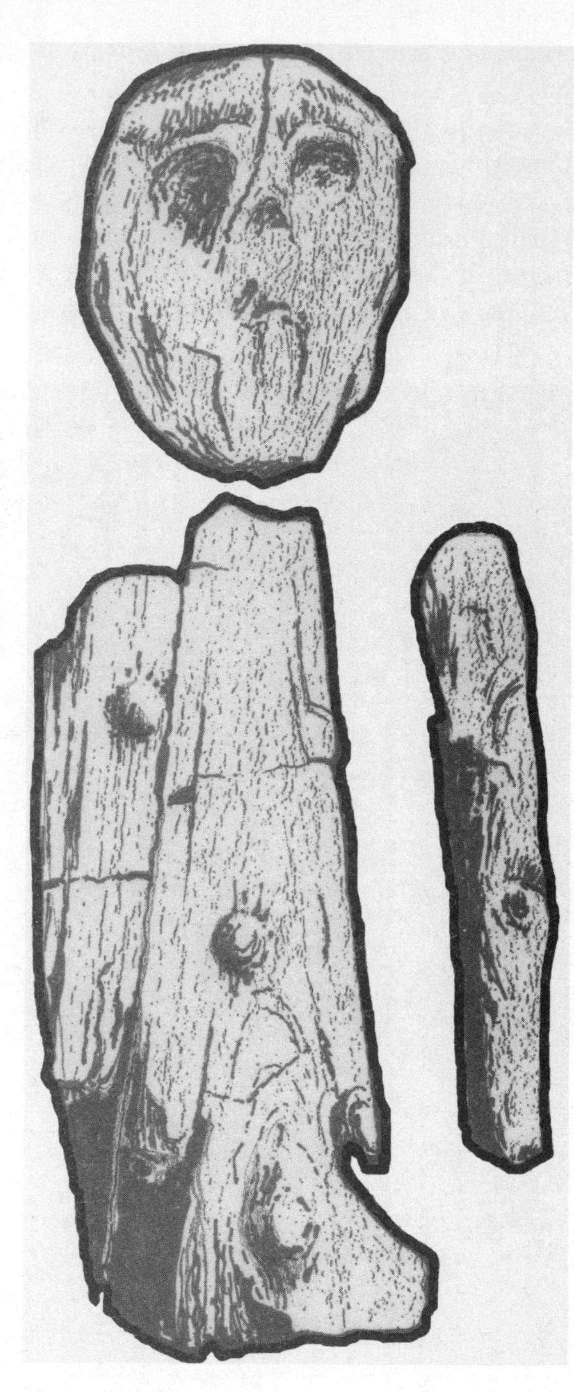

FIGURE 15 Mammoth ivory puppet from an Upper Palaeolithic grave at Brno II (26,000 years ago). (Drawing by the author.)

and the need to practice procedures over and over again must have led to rituals, which are essentially formalized procedures. It may be that little distinction was made between those procedures that we would view as strictly materially causative, and those which, like a genuflection before the altar, are not provably so. Rituals channel rote learning and provide a psychological prop for when things go wrong, as, in an increasingly interlinked and interdependent world, they will.

Curiously, it turns out that as each new engineering, chemical, and mechanical principle that would ultimately lead to modern science was discovered, it tended to generate as a by-product an ever more organized set of superstitions. Although we may never understand all the details from the evidence that survives for our analyses, some general trends are discernible.

One of the most iconic images of the Ice Age is the statuette known as the Venus of Willendorf, carved some 25,000 years ago from limestone and painted in red ocher (figure 14). Alternatively seen as prehistoric mammoth-hunter porn or a celebratory self-depiction by matriarchal, goddess-worshipping women, the naked form has lost none of its impact over the millennia. Large numbers of the little figures are known, variously made from stone, mammoth ivory, and baked clay, in a zone linking southern Russia with the French Pyrenees. Olga Soffer, professor of anthropology at the University of Illinois at Urbana-Champaign, informally comments that because they "have emotionally charged thingies like breasts and buttocks, the Venus figurines have been the subject of more spilled ink than anything I know of.... There are as many opinions on them as there are people in the field."[10] Not just breasts and buttocks, but labia and clitoris as well—but no eyes, nose, or mouth.[11] This relative lack of personal identity is shared with the vast majority of other similar figurines—although some do have schematic eyes and/or mouth, none of the features can be said to approach an attempt at individual portraiture.[12] The emphasis on sexual anatomy will be examined more closely in Chapter 8; here we must focus on the elaborately decorated top of the head. Is it curly hair that we are meant to see there, or a braided or knitted hat?

There is good evidence that fiber arts had become highly elaborated in Upper Paleolithic culture, that it was done by women, and that it produced everything from nets and traps to basic clothing and fashion accessories.[13] If the Venus of Willendorf is understood to be wearing a fabric hat, then her facelessness can be explained by analogy with shop-window mannequins, used as presentation models for fashion options.

As one of the textile conservators connected to Soffer's team put it, the figurines might have been "the prehistoric equivalent of the Sears and Roebuck catalogue."[14]

In this idea of the Venus of Willendorf, Coke really does meet the Ice Age: with figurines seen as the means to promote product, sell a concept, move a specific idea about a useful item or invention out of one mind and into another. The possibility that culture might work like this more generally, by transferring units of information from one mind to another via things, or that particular types of things might spread themselves and become increasingly refined through a kind of cultural reproduction and intergenerational competition is a particular dream of hard-line genetic Darwinists such as Richard Dawkins.[15] At issue is the question of whether there could be definite units of human culture. If there are, and they behave in a Darwinian fashion, then the autonomy of the artificial realm is compromised. System 3 can be reduced to System 2, and ultimately allowing human culture to be fully understood in straightforward biological terms.

Not content in *The Selfish Gene* with atomizing organisms into the little mechanical crawler genes that infest them, each looking out for its own reproductive future, Dawkins suggested that there was a gene-like cultural unit. Dawkins wrote:

> I think Darwinism is too big a theory to be confined to the narrow context of the gene.... I think that a new kind of replicator has recently emerged on this very planet.... It is still in its infancy, still drifting clumsily about in its primeval soup, but already it is achieving evolutionary change at a rate that leaves the old gene panting far behind. The new soup is the soup of human culture. We need a name for the new replicator, a noun that conveys the idea of a unit of cultural transmission... *meme*.... Examples of memes are tunes, ideas, catch-phrases, clothes fashions, ways of making pots or of building arches. Just as genes propagate themselves in the gene pool by leaping from body to body... so memes propagate themselves in the meme pool by leaping from brain to brain in a process which, in the broad sense, can be called imitation.[16]

The idea was that memes were units of culture just as genes were units of biology, and it had a powerful appeal to a large number of people.

It delighted those technologists, psychologists, and behaviorists who, it turned out, had been subconsciously irritated by the endless inability of archaeologists like myself to go beyond describing the way in which culture changes over time to actually explaining, definitively and once and for all, *why*. Here was a simple (or apparently simple) idea for naming cultural units. The successful ones would be copied, and so reproduce and survive in the "meme pool," while the unsuccessful ones would die out. In a way, it could easily have been wholly Panglossian (remembering that Dr. Pangloss in Voltaire's *Candide* insisted that all was for the best in the best of all possible worlds); that is, memes, identified in whatever way in the present, have to be successful, because they are successful.

And so what? Dawkins, having let his new word out of the box, could not get it back in. Unhappy with his original formulation, he nuanced it, making it more consistent but also perhaps vaguer; but memically (one might say) it spread.[17] That is just the problem with what appear to be memes: a meme is not a meme, or, at least, my meme is not Dawkins's meme, as I cannot accept his definition. To me it is a flawed idea, not always wholly wrong, but often incoherent, and unnecessary. It is not, in short, a clearly defined concept, nor one that really solves anything.[18] Dawkins does not like some memes (in his definition) that have proven fitness—the God meme, which would have to be ranked, if we accepted the idea, as one of the most successful ever, is one of those. And he has had to spend the better part of his last two books arguing, essentially against himself, that memes that survive are not necessarily fit in human terms—they are like viruses. So religion and the idea of God are pathologized not as part of our real and authentic cultural DNA but as processing errors—horribly contagious malfunctions akin to cancer.

Dawkins concluded that the computers in which memes live are human brains. Yet these are notoriously complex and not yet well-understood organs. In particular, how our brains respond to, and are reflexively affected by, material things, from natural objects to the complex artifacts that our technology has produced, is only beginning to be systematically investigated.[19]

I have fond memories of one of my first and most significant teachers. One day in 1979, in the middle of a lecture on *The Domestication of the Savage Mind*,[20] the influential book that he had published a couple of years previously, Jack Goody (now emeritus professor Sir Jack Goody)[21] picked up the blackboard eraser with a flourish, pressed it, and turned to the screen. No next slide appeared. Still in mid-sentence, he tried again. Again nothing. Bringing his sentence to a temporary halt with a particularly long *errr* (lectures with Jack Goody are richly punctuated with this sound), he peered out into the projector's dazzle. Spotting the origin of the remote cable (truly remote remotes had not yet been invented), his eyes tracked it as it ran down the middle aisle toward him. Even if we had been unable to see where it went next, we would have known from his head movement that it was looped once round the brass stem of the lectern lamp. Having negotiated this loop, the gaze came to rest on the desk, next to the hand that still held the useless eraser. There lay the black hand control unit. The lecture theater was hushed as a critical comparison was made, reached its resolution, one object was swapped for another, the next slide appeared, and the sentence ambled to its completion.

It would be unfair to describe another's difficulty with words and things without admitting my own trouble negotiating the world of objects while words are floating in my brain. My students also watch bemused in lectures as I try to write on the whiteboard with a canister of spectacle-cleaning fluid—white and cylindrical and with a stubby red cap, just like the whiteboard marker. And my family notice when I confuse the colander for the grater. Both are metallic, shiny, with holes of a certain size. The fact that one is for draining vegetables and the other for grating cheese is a distinction quite far along in my object-processing hierarchy, focused as it often is on the formal qualities of things whose uses no one really understands. The behavior confirms my family in their view that I am impossible to be around when writing, and things are not helped when I use oven gloves to pick up a page of notes lying in the kitchen in the mistaken belief they might be hot (having just burnt myself on a roasting tin). It isn't that I am adrift in the kitchen. I love cooking. But the close choreography that objects require is peculiarly antithetical to writing about it.

It is easier to observe and analyze the problem in others. The board eraser problem arose because the physical world dropped beneath the radar as our professor marshaled his thoughts, leaving inadequate brain power free to deal with the minor misplacement of apparatus. I suspect that something like the limbic system—essentially the reptile brain—was brought into action. While the cerebellum continued to produce a coherent social anthropological argument, old-evolved, prehuman survival routines clicked into play, piggybacking on the higher control centers. This allowed a first-principles, nonverbal, "thing-based" approach to solving the problem: which was that what was being held did not cause the pictures to change. There was a dim perception that this might be because it was not physically connected to the projector. Goody retraced the chain of causality in the way that a raven or a chimpanzee might do it, all the while leaving his verbal machinery free to seek the appropriate verbs, nouns, and modifiers with which to educate us concerning *The Domestication of the Savage Mind.*

With no words in common, the Aboriginal Tasmanians exchanged objects with the first Europeans they encountered—the open-minded, friendly, and avowedly antiracist French researchers. Within minutes, one of the Aborigines had been shot dead and the scene set for a tragic clash of cultures.

I have already, in Chapter 2, described how on first meeting the Tasmanian Aborigines, Marc-Joseph Marion du Fresne,[22] motivated by Rousseau's idea of the noble savage—humans being essentially biological creatures, all equal under their divisive badges of rank, status, gender, and profession—sent two of his sailors ashore naked (supporting the philosophy in the intellectual abstract rather than the corporeal concrete, he kept his officer's uniform on)

The location was Cape Frederick Hendrick (modern-day Marion Bay and North Bay), where Tasman had anchored over a century previously, meeting no one. An elderly Aboriginal man came down onto the beach and presented a burning firebrand. The sailors gave a mirror in return. Marion du Fresne's second-in-command, Duclesmeur, later wrote that the old man was astonished and that "the other savages showed incomprehension as one after the other saw themselves in it." The French were encouraged by this reception, and du Fresne and Duclesmeur followed on: "The spot where we disembarked was dominated by a large rock of which the natives were in possession. However, several of them came down and presented us with fire which we accepted, giving them in our turn some scraps of cloth and some knives."[23] But, as a third rowing boat approached the shore, things turned sour; the Tasmanians launched a shower of spears and stones, wounding both du Fresne and Duclesmeur. The French returned shots, wounding several, and eventually, on a second attempted approach, killing at least one of the Tasmanians. On his return to France, Duclesmeur told Rousseau this, as well as the news that, having left Van Diemen's Land (Tasmania) for Staten Landt (New Zealand), du Fresne had been eaten by Maoris. Rousseau, pondering both events, wondered, "Is it possible that the good children of Nature can really be so wicked?"[24]

This was but one of the unfortunate encounters that, beginning with the best intentions, eroded the Enlightenment belief in the "psychic unity of mankind." That doctrine was borne of resistance to racism and a hatred of the inhumanity of the Atlantic slave trade. Yet Darwin, as fervently antislavery as he was, and a deep believer in humanity as a single (yet, crucially, currently evolving) species, would effectively give his imprimatur to the emerging establishment view that an innate gulf separated the mentality of "savage" and "civilized." The whole of his theory of the evolution of human intelligence was, after all, centrally based on the belief that active selection was under way among more and less intelligent humans; that might just mean individuals but in practice it also meant groups.

It was this resurgent set of beliefs in gross inequality, and a fundamental difference between "advanced" and "primitive" minds that Jack Goody attacked in his book. Goody was promoting the idea that differences in mentality are not so much genetically grounded, but vary between human populations according, particularly, to the form of recording and communications technology available (seeing a massive disjunction in ways of thinking as people cross the threshold into literacy, for example).

I think the lack of comprehension on the beach at Cape Frederick Hendrick was inevitable, given the polar differences in attitude to technology. European technology, developing at an extraordinary pace, was patently frightening... even to Frenchmen. Just a few years after Darwin visited Tasmania on the *Beagle*, Carlyle spoke of an "Age of Machinery" (see epigraph to this chapter), and the following year, the French Marquis de Custine, a well-traveled man from a well-developed European country, candidly admitted (and the meaning, in context, is clearly literal), "in England I feared machines."[25] The transformation in technology that the industrial revolution had wrought on a small island in the Northern Hemisphere can be traced through the statistics on the numbers of English patents taken out in each full decade from the year

1700 through 1849: 22, 38, 89, 56, 82, 92, 205, 294, 477, 647, 924, 1,124, 1,453, 2,453 and 4,581.[26]

If a man like de Custine might fear English machines, what must the Tasmanian Aborigines have thought of the general European level technology that confronted them? The evidence is that the ships alone were objects of terror. With their multi-multi-multi-techno-unit makeup, and their repair, maintenance, and adaptation schedules, they were essentially vast, flexible, multipurpose tools, sustaining a world in microcosm across great stretches of ocean. They must have seemed like UFOs when they first appeared on the horizon, and actually were much closer in logic to modern spacecraft than to the contemporary Tasmanian spear or digging stick. This may well explain why, when Tasman first sailed around more than a third of the island's coastline, past the heartland of at least four tribes, he deemed the land uninhabited. He saw no fire smoke, yet, ordinarily, the Aboriginal tribes would have lit fires whenever they stopped. (In Chapter 8 we will see what happened to another Frenchman, Antoine Raymond Joseph de Bruni d'Entrecasteaux, when he was confronted with an abandoned Tasmanian beach twenty years after du Fresne.)

If European ships could almost be understood, mirrors could not. For a people without even glass (though the Aborigines used natural volcanic obsidian for tools),[27] the idea of a shining screen showing your own face must have seemed too close to whatever they understood as witchcraft.

Giving a gift can be an invitation to change. Technological gifts may be the most provocative and potent of all, as they appear to have no ideology. They do not preach, they simply silently invite use. But in the practice that then emerges, everything may change, and your culture and identity may disappear. Yet gift giving is one of the most universally recognized human traits, and one we share with our primate cousins. There is a distinction to be made, then, between the gift of something old and familiar, or new and fresh but of a known type, and the proffering of a unique object.

In gifts, both newness and antiquity have attractions that the pre-used, shop-worn, or hand-me-down does not. Isaac Babel once wrote that "No one in the world feels the newness of things more strongly than children do. They shudder at its smell, like a hound scenting a hare, experiencing the madness which later, as adults, we call inspiration. And this pure and childish feeling of proprietorship over new things communicated itself to my mother. We spent a month getting used to the pencil case."[28] Sometimes it is true and sometimes not. The magic of the pencil case that Isaac Babel describes has to do with extension of personal history: this is the case into which your own pencils will go, to write your own thoughts. By being new it can become yours and help carry through your intentions on paper in the same way as your fingers are yours and have been trained to write. Its purity is ritual. The child senses the first possibility of controlling a personal destiny.

The gift of an heirloom is different: in this case, the object itself carries the history that you value. Things can go from new to heirloom in seconds, as when Elvis Presley, in the worldwide satellite broadcast *Aloha from Hawaii*, has a series of silk scarves draped round his neck by backstage crew so that he can ostentatiously wipe the sweat from his brow and bestow a saintly relic on a lucky (female) member of the ecstatic audience. Three times the brow is wiped and a kerchief or scarf conferred in an act of communion. The recipients would not have given two cents for a pristine scarf. Now the magic lies entirely in beneficial ritual contagion.

And it is not just faint traces of the King's sweat that can be treasured. Long before the Internet or the telephone, when, despite a rudimentary intercontinental postal service, many people still had to have letters written and read for them, poverty-stricken rural youths in my home county of Norfolk, incentivized to emigrate by sheep-farming concessions in Australia, could pee in a glass bottle before leaving home. Sealed and hung by the back door, the bottle would be consulted by relatives over the years, the condition of the urine indicating whether the person was healthy or ill, alive or dead. Given that many family members were never again in touch, it did not matter if the signal from the urine announced a death, and candles were lit in church to complete the social processes of mourning, out of sync with the realities of physical death on the other side of the world. Goody drew attention to things like this, where the type of communications technology generates a distinctive mentality. And while the urine is not exactly an attractive parting gift, it has something of the spirit of the classic heirloom gift, perceived to have the power to retain connection to the giver.

My elder daughter, just arrived from the pub, has read this over my shoulder and at once placed a drumstick by my left hand. It is signed John Fred Young. The Black Stone Cherry drummer threw it into the crowd at the gig Becca went to, and it fell like magic (of course) into her outstretched hand. The drumstick, once his, is now hers. But it also remains his, and is never fully hers. Ownership is also guardianship. The physical object is her property, but the spirit in it is the force she now guards.

Such a spirit feeds cumulatively as it passes on, growing in history and association, as the pioneering anthropologist Marcel Mauss argued in his influential 1950 classic, *Essai sur le don*—translated into English as *The Gift: The Form and Reason for Exchange in Archaic Societies.*[29] But what Mauss did not quite figure out was how much of what he described could still apply to the modern secular society to which he belonged. This has perhaps been the greatest mistake of the previous generation of anthropologists: seeking to describe the exotic, which they did with commitment and brilliance, they sometimes failed to see that they were uncovering something equally close to home.

The drumstick is now, briefly, with me. I know it is being shown, not given. Like a museum curator, I have been made responsible for it—if Becca gives it away or, in some more distant future, the executors of her will confer it on another, it will be an event dignified by its little history. The drumstick is no longer just one thrown into an anonymous crowd. It has the subsidiary power because "Dad wrote about it." Not that I had a choice. I felt obliged in the same moment it was placed on the desk beside me.

Excavating a cave in the summer of 2009 with my team in the Yorkshire Dales National Park, we came across what may be the earliest traces of human occupation in northern Britain. The find consisted of just a couple of reddish-looking human teeth, but their situation, in a distinctive sediment under a rockfall believed to have been formed by climate change at the end of the Ice Age, means that they may date back as far as 10,000 to 12,000 years ago. In the same deposit was a tiny, drilled marine periwinkle shell of a kind from which late glacial hunters made bead necklaces. These tiny diversions from the natural, in this case the transportation of a shell from the seacoast and its modification along with others to make a personal adornment, have latent force in creating and changing social structures and altering human culture. At the site of Brno II in the Czech Republic an Upper Paleolithic male grave dating to 26,000 years ago was discovered, containing 6,000 similar shells. The beads, strung on a necklace, may have been a present. And once such presents changed, hands it became both easier to leave and harder to forget. Symbolic culture is a form of recording and communications technology. The face of the giver is remembered, even when they are absent, and social relations, forged through such durable tokens, can become very different from those in a naturally bonded primate band.

The grave at Brno contained many other symbolic objects, such as rhinoceros ribs, reindeer antler, pierced disks of stone, bone, and mammoth ivory, and all, like the skeleton, bearing traces of red ocher pigment. But the most striking object—to me the most striking object in all of Upper Paleolithic art—is the puppet: incomplete and badly worn, it survives as just three pieces of mammoth ivory.[30] There is a body on which a belly button, pelvic bump indicating maleness (not quite a penis), and one nipple are clearly visible, and a head with a distinct face drilled up into where it was originally attached, probably with a rod, to the torso (see figure 15). The left arm survives, as do articulation points for the legs, which, like the right arm, are missing. The puppet is the earliest one known from anywhere in the world. Whether or not the person buried was a renowned maker of things—and the broad range of materials in the grave suggests that possibility—the unique puppet figure represents the best empirical data we yet have on a remarkable and far-reaching cognitive development.

Simply put, representative and symbolic art conjured up the idea of divine powers of creation. In particular, the making of an articulate humanoid, which begins a tradition ultimately leading to modern robots, inevitably created a psychological reverberation. Whether or not this was the intention, the question of the force that—or the "who" who—made real humans must now have arisen more sharply than ever before. These, after all, were fully sentient, linguistic artificial apes, mentally capable of being just like us. The related puzzles of death and creation, of what had made the bones in the grave at Brno II cease moving, despite their continued articulation, brought into concrete existence not only the idea of the soul as the absent animating quality, but also, I believe, the idea of some kind of creator god, who made humans, just as humans made puppets.

So the increasing entailment of technology, via necessarily ritualized procedures, leads to ideas about chains of causality in which the material and the immaterial, the appearance of things and their inherent qualities, are bound up. Things begin to be able to stand for other things: soot on a cave wall for an aurochs, a lump of limestone for a woman, a mammoth ivory limb for a human arm. The act of a living human arm trained to use flint craft tools carefully carving a miniature, articulated human arm, from mammoth ivory that had in turn been won from a living mammoth using flint weapons and butchered using flint tools, and whose great curling tusks had been apprehended first not in wild nature, but on an initiation rite deep in a cave, where mammoths were depicted—this is no longer a simple act. It involves many chains of material entailment, and with wide psychological ramifications. Somewhere at the shadowy center of these is the idea of the force of creation, merging the creator and the created. Making the Brno puppet is not just a moment of technological innovation that will lead to modern robotics and the creation of cyborgs.[31] It also marks the moment when its antithesis, anti-science creationism, began its career.

The details of the unfolding of System 3 are so complicated that, over the next few chapters, we shall only be able to scratch the surface. If I and many like-minded thinkers are correct, then the artificial products of humanity have an autonomous logic that unfolds and then folds in on itself, producing unexpected qualities and phenomena. They may initially emerge from the System 2 patterns of biological, Darwinian competition, and be driven by human intentions that are aggressively gene-centered in origin. But they do not come in neat analytical units (even if some of them, sometimes, are meme-like). We glimpse the next level of complexity, as technology begins to come off the leash.

CHAPTER 7

SKEUOMORPHS

Things that try to look like things often do look more like things than things.

—Terry Pratchett, *Wyrd Sisters*[1]

IT MAY BE THAT Terry Pratchett is a fan of the absurdist humorist J. B. Morton (a.k.a. "Beachcomber"), as both are fascinated by the mismatch between appearance and reality—a frequent theme in Morton's fictional news snippets, including my personal favorite: "A man stood on a red-hot egg for forty minutes for a bet at Wivenhoe early yesterday morning. The egg was found to be quite cold, and only painted red-hot."[2] This is a verbal expression of a visual idea about a sensory experience of a potentially unpleasant kind, and the perceptions of the observers, whose mirror neurons (to bring the analysis up to date), triggered through watching the activity, produce a sympathetic reaction of discomfort. This, we are invited to suppose, turns to anger and frustration when the fraud is revealed. The description is funny because of the implied intention behind the painting—to fool other people into thinking an ordeal was being staged for public wonder. The choice of object and trivial scale of the event, along with the precisely specified time, add further amusement. Morton's description is actually true to life, in that this is how a mundane thing, a chicken's egg, becomes a symbolic object.

How things become objects, through a particular kind of separation from the continuous background of the material world, is a central archaeological concern. When I picked up the rock and killed the trout in Oregon, it was not just a question of expedient technology: a thing, briefly, became an object, identified, named, and with purpose, capable of being studied for its good and bad points of function, exchangeable, a potential exemplar for a standard artifact type. When I threw it away, Keith and I stopped looking at it in that way, and it rejoined mute nature. For a while, then, the stone became an agent of action; had it been more recognizable—had we chipped away at it and mimicked the standard shape of an Oregon fish whacker (usually made of wood or heavy plastic)—then others might have borrowed it. Through it, a material intention to catch and kill fish would have manifested itself. That might sound overly complicated, but it is not too much philosophy to say that the emergence of technology was and is intimately connected with the extension of the range of human intentionality. Without a car, Keith and I could not have intended to go fishing that day, given the distance involved; without a stone tool technology, our prehistoric ancestors could not have had the intention to kill big game, or make baby slings.

In his seminal work, *Art and Agency*, the anthropologist Alfred Gell pondered his own, genuinely hot, breakfast-time boiled egg and asked:

> What has caused this egg to be boiled? Clearly, there are two quite different answers to this—(i) because it was heated in a saucepan of water over a gas-flame, or (ii) because I, off my own bat, chose to bestir myself, take this egg from its box, fill the saucepan, light the gas, and boil the egg, because I wanted breakfast. From any practical point of view, type-(ii) "causes" of eggs being boiled are infinitely more salient than type-(i) causes. If there were no breakfast-desiring agents like me about, there would be no hens' eggs (except in the South-East Asian jungle), no saucepans, no gas appliances, and the whole egg-boiling phenomenon would never transpire and never need to be physically explained. So, whatever the verdict of physics, the real *causal explanation* for why there are any boiled eggs is that I, and other breakfasters, *intend* that boiled eggs should exist.[3]

Gell accurately described the entailments of modern technology, and implies how the existence of objects, such as saucepans, not just allows actions but suggests them. The ability of objects to suggest things in this way has allowed the development of special features of objects, and special types of objects, where the function is more to suggest than to deliver. An example would be a fake-fur leopard-skin coat, lacking the original insulating qualities of the fur, but imbued with other qualities, such as a capacity for complex social signaling. Such an object, in archaeological parlance, is a *skeuomorph*, a classic manifestation of technology as it leaves behind the realm of natural things.

Skeuomorphs are all around us. Simply put, they are carryovers from an older technology or way of doing things that had value, and are retained as a semblance, and expectation. Characteristic of changes in technology, they confer a kind of luster. The technological reason for the feature has gone, but you expect it—it completes the object. Open a wine bottle and pour out the wine. Notice that the bottom is dented-in, in a shape known in France as *le voleur* ("the thief"), because without it there would be more wine. When wine bottles were blown, there was no alternative: the molten glass bubbled out like a long balloon with a rounded end; this base was then flipped inside out as the bottle was set down to cool, producing the level circumferential basal ring that would allow the bottle to stand upright.

This was in fact the easiest way to make a flat-standing base, a trick known to metalsmiths for millennia. In Bronze Age Europe there were no glass wine bottles, but there was wine, some of it drunk from gold cups like those found in a hoard, along with a great golden serving bowl, at Vulchetrun in Bulgaria—cups with "omphalos" bases. *Omphalos* is the ancient Greek word for the navel—and also for a beehive—and describes a hemispheric form, like half an orange placed cut-side down. Goldsmiths, making a cup from a flat sheet of metal, would first hammer it into a bowl shape, and then, to stop it wobbling when it was set down, would neatly flip in the bottom, making a little half-orange shape in the base. The point where this shape folds is where contact with the table is made.

Wine bottles are not made like this today; they are mass-produced in molds, like the Mesopotamian bevel-rimmed bowls of ancient Sumeria, and they do not need to incorporate *le voleur* in their design, as this was a by-product of a superseded technique. But most still do: the dent in the base has become a skeuomorph. *Le voleur* remains because we expect it to be there.

The Bronze Age specialist Andrew Sherratt and I had a shared love of skeuomorphs, and used them to trace what Andrew called "invisible flows" in prehistory—the trajectory of technological transfer and know-how not directly preserved on archaeological sites, but which can be inferred from skeuomorphs.[4] We were particularly interested in the idea that knowledge of metal, particularly precious gold, had filtered into northern Europe much earlier than supposed, during the late Stone Age farming period known as the Neolithic. There was already a clue when Ötzi the Ice Man was discovered. Dating to around 3,300 B.C., he belonged to an Alpine culture considered Neolithic because the burials from this period in the Alps contain only stone tools. But Ötzi had a copper axe with him, bound with sinew to a yew haft. Such axes are found, as rare objects, in the contemporary cemeteries of the Remedello culture in lowland Italy, on which basis that culture had been deemed more technologically advanced than the Alpine one to which the Ice Man belonged. Now a suspicion began to grow that this was already a standard item of gear in the mountains, but that it would be passed down at death, being too valuable and technologically advanced to be placed in anyone's grave. Because Ötzi died accidentally and was covered swiftly by ice, he was found with objects he literally wouldn't have been seen dead with.

Andrew and I set out to track down metalwork by identifying, in skeuomorphs, the rumor of its passing. A classic example of what we were looking for came from the Neolithic burial chamber of Oldendorf II. Dating from close to the time of Ötzi's death, this site, in Lower Saxony, Germany, was even farther from the shores of the Mediterranean. The grave contained two pottery cups that not only had omphalos bases, just like metal ones—and an omphalos base has no logical technological place in the ceramic arts, making the base more difficult, not easier, to form—but also the appearance of a rivet shape where the handle was "attached" (which, being clay, it obviously did not need).[5] Other things about these cups were skeuomorphic, from the ribbon-like arching "strap-handle," appropriate to sheet metal but a fragile liability in clay, to the sharp profile that looked just like that of a gold cup.

Andrew and I disagreed about the implications. He tended to think that the Oldendorf cups were made by someone who aspired to ownership of actual golden cups from the sunny south and, having no direct access, had copies made. I thought the skeuomorphs were too formally specific to have arisen unless the potter had had real metal exemplars to copy, and opted for an Ötzi-style explanation: golden vessels existed, but cheap proxies were all that were spared to go in the grave. If the gods got the idea, then the dead warrior might be provided with the real thing in the afterlife.

This is typical of skeuomorphs, and what makes them so fascinating and problematic. They establish the presence of the idea of metal artifacts over a wider range than archaeologists are able directly to observe, but their specific meaning is harder to establish. Clearly, as in the case of a fake fur coat, social meaning can be complex: such coats made their appearance as cheap versions of real furs for an aspiring middle class who wanted to mimic the possession of real wealth; later, as ecological concerns over endangered species came to the fore, wearing fake fur became a moral statement, as well as having knowing, kitsch, and ironic references to the past—"retro-chic." Synthetic fur manufacturers, having at first striven to make the fake material look as real as possible, are now careful that it actually does *not* look like the real thing.

It can be hard to distinguish between the real and the fake, and this difficulty extends beyond the world of durable artifacts, to food and even emotion. The traditional Lower Austrian *Traueressen* is a grieving meal of simple boiled beef, accompanied by raw grated horseradish. Calculated to make the sinuses water and perhaps even draw tears, *Traueressen* genuinely contributes to an ambience of solemnity by encouraging the physiological responses we associate with grief; the specificity of the dish, like any traditional dish at a ceremony or celebration, helps maintain consistency of mood. The flavors and appearance are far from the famous golden Wiener schnitzel cutlet. Yet both dishes involve complex illusions.

It was said of the later Roman emperors that they ate their meat encased in gold leaf. Gleaming, luxurious, inert, and edible (if not strictly digestible), it was the sort of fare reserved for the super rich. It was later emulated by the medieval bourgeoisie of Milan, who replaced the gold leaf with golden bread crumbs. This *cotoletto milanese* so impressed Field Marshal Radetzky that he introduced it to Vienna, and by the end of the nineteenth century, it had become an Austrian national dish.[6] Eventually the gold was forgotten, and Wiener schnitzel became exactly what it is.

Is *Traueressen* a deceit? According to the philosopher Ludwig Wittgenstein, human emotional states are communicative and embedded within their mode of expression—that is (in our culture), grief cannot be detached from weeping; weeping is not a symptom of grief, rather it is one of the things grief *is*. Shakespeare observes grief in many forms, showing how grief for another is often really grief for ourselves. This is true, for instance, when Hamlet approaches the grave, newly dug for Ophelia, who he does not know is dead, and the sexton points out the skull of Yorick, the kind jester of Hamlet's childhood.

As I watched a recent production, with David Tennant as Hamlet, something seemed wrong with Yorick's skull.[7] Properly speaking, the skull was just a cranium. The mandible (lower jaw) had become separated (as it likely would in twenty-three years of interment), the eye sockets were stained, the nasal cavity was ragged, the teeth were irregular and covered with convincing grave-bottom silt. The odd thing was the horizontal line across the brow. The skull was either an anatomical specimen sawn open and stuck back together, or a plaster cast with a clumsy casting seam, or a plastic skull of the sort where the top opens to allow a view inside.

I had seen dodgy skulls in Hamlet before, but I knew that, for the stage production on which this filmed version was based, the Royal Shakespeare Company had used the skull of André Tchaikowsky (Polish composer, pianist, Holocaust survivor, and Shakespeare fan). Unimpressed by the props in the 1979 production, he bequeathed his skull to the company. After Tchaikowsky's untimely death in 1982, his skull was used periodically in rehearsals, but not in public until 2008, at Stratford—and then covertly. News leaked out, and when the production transferred to London's West End, the RSC announced they would cease using it, to avoid distracting audiences.[8]

Watching Hamlet on television, I stopped concentrating on the fine acting and watched the skull. And as David Tennant lamented Yorick, the double parallel striation could be seen close-up. I suspected the BBC was trying to reassure viewers that this was indeed a piece of plastic. I began to feel patronized and slightly angry that—in a drama where nine out of ten main characters die horribly—anyone should presume to shelter me from the sight of the physical remains of a distinguished artist.

The truth turned out to be more complicated. It appears that the skull used was Tchaikowsky's, and had been all along, even in London. While giving no real detail about procedures, the website www.andretchaikowsky.com implies that the brain was removed, and shows a stained skull atop the dead composer's sheet music: the whole braincase appears to have been sawn round with a professional skull saw. The play's director issued a statement that a replica was being used because he was worried that audiences would come for that spectacle. They would not focus on the meaning of the grave-digging scene if they felt they were watching the skull of a real pianist rather than a fictional jester. Pratchett's mind-warping *Discworld* observation perhaps sums it up: "Things that try to look like things often do look more like things than things." Or should that be the other way around?

When I first saw Antony Gormley's *Another Place*, the figures were dispersed between the seawall and the water's edge, fully exposed and waiting to be re-covered at high tide (see figure 16). We could make out perhaps a dozen of them, roughly equal to the number of living visitors on the beach, who ran their hands over them, walked among them, or posed, soft hand in metal hand, for a photo.[9] Each life-size figure weighs about 1,433 pounds—ten living humans. Cast from the same original mold, and welded in standard fashion, the hundred sculptures that dot the shore for two miles, extending half a mile out to sea, have acquired history and personality at a micro-level. Weed

FIGURE 16 *Another Place*: image of Antony Gormley's art installation at Crosby, Merseyside, England. (Photo courtesy the author.)

covered, or rusted to a deep red patina, some are buried up to their thighs in sand. Others stand in shallow pools, or have had their anchor plates jaggedly exposed. Many display penises and bolt-like nipples splashed with white or yellow paint by irreverent local youth. Fiercely adorned by nature and culture, the blank mannequin faces show no emotion. Yet the stance of the figures, legs slightly apart, arms tensed away from their sides, suggests at least the contemplation of action. Gormley's sculpture is tangibly a masterpiece. Seeing it myself in the wake of recent visits to both the *Venus de Milo* and the *Mona Lisa* in the Louvre, and the polychrome Paleolithic paintings at the cave of Peche Merle (dappled horses surrounded by soot-stenciled outlines of human hands set among the stalactites), I found the figures of *Another Place* an equally demanding and original statement of what it is to be human. But if anything could be said to be a meme, that proposed reproducible unit of culture akin to a gene, this standard-template reproduction of the artist's own body in metal should surely qualify. In practice, the argument might go like this:

Antony Gormley's genetic inheritance, or genotype, gave him particular characteristics. He learned particular motor skills and hand-eye coordination and became an artist. These inputs, chosen and unchosen, produced the Gormley phenotype—the concrete physical appearance with its underlying attributes. The same Gormley genotype raised in a different setting, with different diet and habitual movements, would look different. In the Roman Empire, born to slaves and sent down the mines, he might well have become a phenotype with bandy legs from rickets, and a hunched form and massive shoulder muscles from working underground. If this mine-slave Gormley had ended up in Pompeii in A.D. 79 and been buried alive in volcanic ash, the body-shaped void around the bones, cast in plaster like those on display in the museum in Naples, would not resemble *Another Place* Gormley.

So much for the biology. The artist has reproduced himself in iron in an installation by the sea. Although the artwork appears unique, each of the one hundred Gormley figures is cast from the same mold. These are artistic clones, or as Dawkins would say, memes of one another. They are also unoriginal in another way. The imagined Roman Gormley clone would have been very familiar with cast-metal statuary in public places, so the installation at *Another Place* is doubly memic: the idea—the "meme"—for an artistic representation of a human being has been around for a very long time, copied down the generations, from before ancient Rome, back to the time of the Brno puppet and the Venus of Willendorf. So (according to meme theory) everything in the world of artifact culture is just another version of an underlying template, reproduced by a process of copying. Gormley has merely copied Alexandros of Antioch, who, in turn, falling in line with some long chain of intervening artistic geniuses, has copied the master or mistress of Willendorf.

But is *Another Place* really just another version of the *Venus de Milo* meme, itself a version of the Venus of Willendorf meme? Is Gormley's work an unoriginal reiteration?

There is a second, more fundamental problem with the meme. Most meme theorists seem to assume that the meme, whether it is in the head or in the real world, is a fixed thing, the thing that gets copied. For example, a chair seems like a simple sort of meme, being something you can sit on. But then again, a log is something that you can sit on, as is a sofa. Even people can be used as chairs (as *Moby Dick*'s cannibal harpooner Queequeg says his father's slaves were[10]). So is a chair the same meme as the idea of "sitting down"? That is, what is the essential blueprint for the chair idea? The best thing to do at this point might be to examine a chair and see which features must be copied for the chair meme to be successful: a seat, a back, legs, and maybe arms (an armchair). But some have a single pedestal, and others fold down from the wall or are built in. Essentially a chair needs a seat to sit on, like a sofa or a bench. But then how does it differ? Perhaps by being designed for one person, like a barstool (which is a backless kind of chair).

Having got this far, it would be useful to clarify whether a chair is a seat and if so, what features are both necessary and sufficient to define one. The terms *necessary* and *sufficient* may be technical sounding, but it is clear that if you want to design a chair, you need to know what it needs, and how its requirements differ from those of a sofa or a barstool. So a chair needs to support a person in a seated position off the ground. Chairs in diners are often fixed next to one another in an array of four, like the fold-out chairs of our old integral camping table; transposed to a railway station waiting room, an identical seating technology without the table in front and without reciprocal facing chairs becomes a benched row of seats, or simply a bench with individual hollows for seating. The point at which the technological module stops being a chair, singular, is now unclear. Perhaps there is not so much a chair meme as a seat meme, which is rather more general. Or, on the other hand, perhaps there is a much more specific set of memes for armchair, dining chair, highchair, executive chair, picnic chair, folding chair, designer chair, and so on. But if there is a meme for a designer chair, designers would be suing each other for breach of copyright rather than competing with one another and approving each other's innovations while jealously guarding their next big idea. Memes are supposed to be about replication, not innovation. The distinctively human usefulness of the chair meme is that once it has been invented, you don't have to think about design anymore... do you? This was the great failing of Soviet design—the assumption that basic utilitarian objects could be defined and cloned, without the essential spark of human creativity.

It turns out that there is no rule that defines a chair. The idea we have of a chair *in practice* is a fuzzy intersect of attributes, none of which is in and of itself both sufficient and necessary for inclusion in the category. Its *necessary* attributes are not sufficiently distinctive to specify a chair—being a place to sit, of a certain shape and structure, entails overlaps with other items of seating technology, and its only *sufficient* characteristic is linguistic and symbolic, namely that we should agree—whatever its number of legs, regardless of the presence or absence of arms, whether it is made of wood, or plastic, folding or built-in—to call it a chair, because, like Gell's boiled breakfast egg, we intend it for that purpose.

In the technical summary it goes like this: objects of the kind where no single attribute is at once sufficient and necessary for the classification are known as *polythetic*, and are typical of System 3; by contrast, System 1 entities (such as a proton or an electron) are generally considered *monothetic* as you can firmly define what it is you mean and how it differs in its characteristics from another entity by pointing to the diagnostic feature that it must have, and if it has it, then that is what it is. Despite some debate, System 2, or biological, entities are broadly understandable monothetically as well.[11] The polythetic nature of artifacts is the logical classificatory reason why the meme concept cannot apply to them, except in very approximate terms.[12]

Things rule us: polythetically defined, produced as objects through human intentionality, deceptive in their skeuomorphism, their slippery and redefinable associations. That such complications existed was something that began to dawn on me as a student after an embarrassing incident in a Viennese toilet. As many authors have discovered, toilets bring the material achievements of civilization—doors, locks, ceramics, plumbing—sharply up against urgent biological reality.

As a young research student I often excavated in Austria, particularly with my great friend, the Paleolithic specialist Gerhard Trnka. On this occasion Gerhard was pottering around his office at the Institut für Ur- und Frühgeschichte, misting his beloved cacti with the pump-action water spray that a day previously had damped down the trench sides as new strata revealed themselves (from site tool to water spray, artifacts changing name and role according to context and intention). I needed freshening up too—I had finally gained an audience with the elusive emeritus professor Richard Pittioni to discuss a controversial and mysterious silver cauldron about which the professor had, many years previously, written an entire book.[13] I had gleaned that he was old-school formal, and I had managed a bow tie, jacket, and relatively clean jeans. Punctuality was essential, but with the previous night's alcohol flushing through me, a quick visit to the toilet seemed wise.

The Institute's gents' toilet did not have the usual row of urinals, just five cubicles. All were empty, and I was standing in one, when I became aware of a janitor in a white coat standing immediately behind me. He asked with palpable irritation if I thought it was acceptable to use the toilet with the door open. I shrugged and replied defensively (which regrettably means rudely), "Ja. Schon. Und?" (which might translate as "Yeah. Sure. What of it?"). The little bullet-headed cleaner stood his ground in what was clearly *his* toilet, his white mustache bristling, and ordered me not to do it again. Back in Gerhard's office, I expressed my disgust at the *Kloputzer*. What sort of "bog cleaner"? Gerhard asked. There was a cleaning lady who came later; perhaps I might not be aware that Professor Pittioni was of the generation of university professors who religiously wore a white lab coat at work?

It would have been worse now not to keep the appointment. At my knock, I was called into the dim, book-lined office. The diminutive, balding Professor Pittioni sat at his desk, academic publications spread out in front of him under the focused glare of a metal angle-lamp. He rose to shake hands, the white acid-resistant rubber buttons of his lab coat starkly lit. The initial tension passed, but he made no acknowledgment of our first, accidental meeting. He was civil yet intellectually unilluminating, his views on the cauldron having apparently fossilized on the completion of the work I had read. From time to time his eyes gleamed with what could have been amusement, but there was no corresponding smile. I realized I would never be allowed to apologize. This was my punishment (which Pittioni, I think, enjoyed).

Pittioni never worked in a laboratory, whereas few lab scientists in British archaeology wore white coats anymore. Looking at animal bones under binocular microscopes, gluing pottery back together, or sorting carbonized seed residues does not demand much more than practical everyday clothing. The white lab coat was worn only when special cleanliness or protection was needed—but in Vienna, it was a mark of status, distinguishing Pittioni as a professor. It had its origin in the white togas of the Greeks, a mark of citizen status, and thus of cultivated learning. After the fall of Rome, white robes symbolized early Christianity and were worn—long and like a skirt—by an evangelizing priesthood. Many of the ancient Scythian and early German pagan priests had been cross-dressers.[14] The long-robed priest in white at once personified the Roman authority of the papacy and the gender ambiguities of the true shaman. Celibate, he was neither a real man nor a real woman and could minister to both.

Perhaps something of that ambiguous purity recommended the garb to the medics of the later nineteenth century. White began to seem both more scientific and more hopeful than the funereal black that had typically been worn in the morgues and while visiting the dying. Chemists, in particular, favored protective plain clothing. White made scorches and stains easier to spot, and new industrial chemicals for laundering allowed white to be worn more regularly in practical contexts.

Doctors of the nineteenth century were also seeking to acquire the transcendent authority of priests. Darwin had deposed the Judeo-Christian God of immediate human creation, suggesting a long-drawn-out process of chance and necessity instead of the blinding light of a divine six-day creation. White, seen as radiant and holy, reminiscent of standard depictions of saints and worn by the pope, was a symbol of compassion. It was clean and purgative, with important connotations of separation. The white coat raised and set important people apart within a special environment where dirt (and thus uncertainty) was banished, and specialists were visually separate from the people they treated.

In Vienna, with the founding of psychiatry by Sigmund Freud, the idea of the analyst as savior perhaps prompted adoption of the white coat beyond the ward and the lab. Its use spread out among art historians and curators, geologists, paleontologists, and so on, signaling seriousness of purpose, vocation, and devotion to truth. The image of Socrates and Plato with white togas floating in the blue air of the Aegean returned to the neoclassical marble halls of European academies of learning. Eventually, the symbol extended even into pornography: the Swedish hard core movies of the late 1960s were termed "white coaters,"[15] as they managed to avoid the censor in Britain and the United States by being introduced by white-robed sexologists—apparently earnest, and certainly only covertly enthusiastic, men who justified the graphic content as significantly educational.

The way in which clothing is perceived may shift sharply across cultural boundaries, as Anton Chekhov discovered in 1890 when he visited Sakhalin Island, populated by Russian convicts and native Gilyaks. "The Gilyak's clothing has been adapted to the cold, damp and rapidly changing climate. In the summer he wears a shirt of blue nankeen or daba cloth with trousers of the same material. Over his back, as insurance against changing weather, he wears either a coat or a jacket made of seal or dog fur. He puts on fur boots. In winter he wears fur trousers. All this warm clothing is cut and sewn so as not to impede his deft and quick movements while hunting or while riding with his dogs. Sometimes, in order to be in fashion, he wears convict overalls."[16]

Between humans, the dynamics of first contact are crucial, the stakes often high, and the scope for misunderstanding immense. In unfamiliar surroundings, we are set adrift. Shorn of the fully human identity that our touchstone objects silently build for us, we become unsure not only what to think, but how. The pioneer social anthropologists often experienced this on their return from months among the tribal peoples of the southern Pacific, and it made them genuinely, physically ill. They called it "culture shock." Since my culture shock in the toilet with Professor Pittioni, I have come to believe that things do not just alter the way we are on the surface; they are actually what has allowed us to become human.

The incident with Pittioni may have been unique (he passed away many years ago, and white lab coats have by now virtually died out even among German-speaking professors), but it has a universal relevance. Unlike animals, we do not "just" urinate or defecate. Like everything else we do, we do it in ways that have meaning, with requirements and limits of acceptability. These apply very particularly to times and places, and to micro-contexts within times and places according to who you are, which sex, what age, which religion, and so on. In modern Swedish restaurants, toilets are unisex, as they are in most English or American private homes; on the other hand, in parts of the Islamic world, toilets are exclusively male and female within the home (and are carefully oriented in relation to Mecca).

As the British archaeologist Matthew Johnson has shown in his studies of the form and meaning of built spaces, if I had been invited to a meeting with Pittioni in a medieval castle, I might well have found myself sitting alongside my host and several others on one of a row of toilet holes overhanging the battlements. We would have held our discussion of the mysterious silver cauldron while unmysteriously adding to the vast heap of steaming excrement below. According to Matthew, the castle excrement heap would have been a display item—a temporary, mutable art installation with shared yet proudly exclusive authorship.[17] This status symbol at the base of the wall showed both visitors and the peasantry just how generous were the portions at the lord's feast, and how numerous his well-fed friends.

Anthropologically speaking, what went wrong with Pittioni and me occurred in the overlap between the *etic*—the shared understandings of humans as a single species—and the *emic*—the fine-tuned cultural categories and routines that distinguish us from one another by sex, tribe, nationhood, age, profession, and so on. Etic things are easily translated between languages. Any phrasebook will tell you how to ask the whereabouts of the toilet. But *how* you should go is an emic matter, full of internal meanings distinguishing insiders from outsiders, betraying fine shades of snobbery, and laying traps for the unwary. The how cannot easily be learned through translating words in the abstract; it must be understood by observing people's interactions with things. The castle battlements and toilet holes physically brought into being one way of life, one way of understanding hospitality and wealth, while the doored cubicles of the archaeology institute toilet cubicles brought forth another form of existence.

Twenty years after the initial and ill-fated landing of naked French sailors on the shores of Tasmania, Antoine Raymond Joseph de Bruni d'Entrecasteaux, with his ships *La Recherche* and *L'Espérance*, spent a month in what came to be called Recherche Bay. At first he saw no one, despite finding an artificial shelter near the beach, along with some kelp water bags and finely woven bark baskets. Nine months later, when he returned to the same anchorage, the Lyluequonny people came out to welcome him.[18]

The Aborigines explained that the ships had scared them first time around. Now, over an intense three-week period, the two small communities established a rapport. They exchanged songs, dances, and culinary tips. D'Entrecasteaux's men wore clothes most of the time, and the Lyluequonny were fascinated by these, and even more by the fact that the French sailors appeared to all be men. The Aboriginal men and women operated together, often with separate roles, but always as a unit. To them, it was inconceivable that what appeared as a mobile community similar to their own, albeit shipborne, could exist without females for its continuation. A better explanation had to be that these "French" were not what they seemed. Suspecting that their clothes served only to disguise a mixture of sexes, they cajoled the officers and crew into lining up on the beach, dropping their breeches, and allowing their penises to be inspected.

An officer on *L'Espérance*, Jacques-Malo La Motte du Portail, later remarked in a letter that the Aborigines had examined the steward of *La Recherche*, "they would have come across what they wished to find."[19] The Tasmanians had no practice in accounting for people in large groups, and they had trouble distinguishing members of the crew from one another: the French seemed to dress alike, and their pale faces were collectively alien-looking. Thus La Motte du Portail noticed something that was overlooked by the Tasmanians: the steward of *L'Espérance* had hidden during the genital examination.

Although their suspicions were aroused by his absence, the French crew continued to respect Louis the steward as a man. In this role he had received arm wounds in an easily avoidable duel with an insubordinate assistant pilot (displaying virility, perhaps, through both bravery and appropriately gender-specific stupidity). Another identity was not revealed until some two years later, when the steward died of dysentery, and the expedition's surgeon signed a death certificate for one Marie-Louise Victoire Girardin. Honored in 2005 as the first European female to reach Tasmania—"Our daring drag king unveiled," as the *Sunday Tasmanian* had it[20]—Girardin was the daughter of a royal gardener, had married young, was widowed, bore an illegitimate child in the early days of the French Revolution, and fled France in disguise.

Tasmania before the Europeans was a coherent cultural system. Its tragedy was that, through reverse entailment, it had come to look natural. With the Tasmanians, what you saw was what you got: there were no drag kings or transvestites. Material culture was so pared down that sex almost *was* gender. For the Europeans, physical aids had long had the potential to level differences between the sexes. Yet, as it became possible for males and females to exchange roles through "equalizing technologies," whether men could now bottle-feed babies or women could shoot physically stronger enemies with guns, clothing codes for men and women became, if anything, more conventionalized. The more canonical they were, the more they could hide as well as express biological reality, artifice superseding nature.

During a recent (successful) campaign to stop commercial logging in Recherche Bay, the French scientific visit was held up as the intercultural contact that might have been: as the rights activist Colin Hughes put it, "The French treated us like people, not animals.... It's a damn shame everyone didn't do the same."[21] But research parties unfazed by challenging cultural phenomena, whether naked tribespeople or cross-dressed fugitives, became scarce. Ex-cons settling down to extract a profit were the new reality—people from a culture where asserting an unambiguous identity was ingrained. What counted was size and type of houses, mode of transport, number of servants, the sort of clothes and jewels one could afford. The settlers were people with no interest in seeing humanity in those who existed as laxly as the Tasmanians appeared to do. As space was made for sheep and cattle ranching, logging and mineral extraction, dancing on the beach and penis inspections gave way to terrorization and murder. Hobart arose as a city, with a city hall, cathedral, police station, law court, port authority, prison, school, library, gallery, and museum. Paul de Strzelecki saw things cutting both ways, noting that the indigenous Tasmanian way of life represented a valuable kind of freedom: "the maladies arising from the... artificial state of society, are unknown to [them]."[22] For the Europeans, everything had long since been revolutionized by an artificial state, psychologically as well as functionally.

The conventions of European clothing were only partly about patrolling the boundaries of sex, making those parts of the physical body that were discreetly hidden visible again by proxy. Gendered clothing has helped give other differences—other inequalities—some sort of illusion of natural difference, by embodying a conceit that the forms of clothes revealed the nature of the individual. This used the touchstone of gendered clothing (which did hide or enhance an underlying biological reality) to extend signaling far beyond the man/woman distinction.

It is less a question of dressing different classes and statuses of person differently to identify them more easily, and more about creating the very idea of such categories, producing an effect of naturalness and inalienability. Different social roles could then appear as basic as male and female were in reproductive terms. Queen and priest, soldier and sailor, doctor, washerwoman, librarian, joiner, shepherdess, policeman, and prostitute, all wore more or less gendered clothing, differentiated by many other conventions and rules, enforced by the laws of sumptuary, insignia of rank, and symbols of power.

Such games were not in the Tasmanians' psychological or (and it becomes the same thing if you follow the argument of this book) technological repertoire. Recognizing few social differences in their small bands, they were unused to treating people with the grades of deference to position that living in a city, or a modern state, would require, even if they could grasp the concept of living indoors in a fixed abode. How could they ever become tax-paying subjects? As de Strzelecki put it, "the attempts to civilize and Christianize the Aborigines, from which the preservation and elevation of their race was expected to result, *have utterly failed*" (his italics).[23]

The Tasmanians' naked egalitarianism may seem more admirable to us today than the straitjackets of class, sex, and status that then governed European lives. And for Louis/Louise, who belonged biologically to System 2, the reality had become defined by System 3: she could continue to sail with the tacit knowledge of the French crew only because she observed gender decorum. She was free to sail *as a man*—that is, dressed as a man—only by respecting a gender grammar which trumped the underlying biological reality. But which was the reality? Beachcomber's red-hot egg at Wivenhoe was not really red hot, but it was treated as if it was, and that, at least for a time, was all that really mattered: as the centerpiece of a red-hot egg spectacle, it functioned like a red-hot egg.

Drag is a skeuomorphic subversion of the naturalization of differences that codes of gendered clothing bring into being. It undermines the logic of outer codes, and for this reason it is disapproved of especially strongly in theocratic and autocratic states, because it is no longer clear if people are behaving appropriately to what is fondly regarded as their natures. It should come as no surprise that the country with the highest published rate of sex change ("gender reassignment") operations is modern Iran.[24] Surgical conforming of the physical body to make an ingrained behavior that is considered deviant (especially that of homosexual males taking a perceived female role) is a classic instance of System 3 projecting itself back into System 2 rather than temporarily and unstably masking it. The artificial reconstruction of the body seems to be seen by Iranian sex-change surgeons as an act that returns a sense of natural order to the disordered body. In fact it is artificially designed to project and reinforce a specific cultural idea of acceptable social relations.

Looking at the famous picture of the so-called last Tasmanians, Truganini and her cosurvivors, we see four people, all outrageously uncomfortable in full-blown mid-Victorian outfits.[25] Going by the dress codes, we discern three women and a man. Their expressions, unhappy and put-upon as they certainly are, suggest something else: a blank disbelief that they have joined a human society in which the natural could have been so entirely masked and screened. If there is truth in Rhys Jones's "slow strangulation of the mind," it kicked in here, in these stultifying clothes and staterooms, in these ten seconds, as light, reflected from their faces, was refracted through a ground lens, and the collodion process bound shades of silver iodide onto glass. The photographic image suggests that its subjects would rather have been cracking fresh lobster by a roaring fire under the stars, naked, greased up, and off camera.

From this perspective, perhaps Rhys Jones's other comment, about the Tasmanians' minimalist native technology indicating a "squeezing of the intellect,"can be twisted around.[26] Their mental horizons *were* limited by, defined by, and created through their material culture. This is true for everyone: it is the human condition to be more than just flesh and blood. The name Truganini, in the language of the Nuenone people of Alonna-Lunawanna (now South Bruny Island), meant a person who makes tools.[27] Our minds and wills extend through artifacts—stone blades, ships of discovery, men's and women's clothing, photographs. Most of these systems of material culture we inherit. They are familiar precisely because we did not invent them—they are just "how the world is." For us, they have become so complex that we really do not know how they work anymore. We are increasingly unsure where we end and they begin.

Transvestism is an entailed technology too. It took Marie-Louise Victoire Girardin out of a world of rouge and powder and restrictive clothing (inhabited, pre-guillotine, by aristocratic men as well as women). Her disguise revealed a new side to her—him—and enabled physical engagement as part of a male crew on the high seas.

My friend Ben Wheatley has sailed aboard the Tasmanian ship *Windeward Bound*, registered in Hobart and captained by Sarah Parry. Sarah, the partner of mother-of-three Jennie Kay, is a Vietnam veteran and former Royal Australian Navy clearance diver who used to be known as Brian. Parry says, "I tend to see myself as a third gender. I live as a woman. I'm known as a woman, legally I am. However I'm able to use all those things that I learned as a male and put them to good use in my life now. I don't reject my previous life."[28] Although her genome remains the same, through modern medical know-how, Parry's body has been remodeled away from being male toward being female. What eighteenth-century Tasmanians and European sailors would have made of this can only be guessed at. What would Rousseau or Darwin have thought? Contrary to their expectations, what Parry's experience, as well as those of Girardin and Truganini, signals is that being human, in any terms, long ago became something more than natural.

Chapter 8

SCREEN CULTURE

Technology is not politics pursued by other means; it is politics constructed by technological means.

—Brian Pfaffenberger, *Technological Dramas*[1]

"KILL LOOTERS, URGES ARCHAEOLOGIST" ran the headline. Professor Elizabeth Stone, head of archaeology at Stony Brook University, New York, addressing a 2003 British Museum conference on the archaeology of the Iraq region, was reported as saying, "I think you have got to kill some people to stop this....I would like to see some helicopters flying over these sites, and some bullets fired at the looters."[2] The idea disturbed me. Surely human life is more precious than old bits and pieces found in the ground, even if they belong to the world's first civilization?

Stone was not complaining about the theft of things like BRBs—bevel-rimmed bowls. Scholars themselves may care little enough about those, as one researcher candidly admits: "Unhealthy archaeologists in the field...find that BRBs make handy ashtrays or soap dishes (the veiled contempt for these hideous containers is tempered by their inherent usefulness)."[3] Rather, Stone was particularly upset that, in the search for unbroken artifacts to sell on the art market, many of the incomplete records of the earliest writing in the world were being thrown away and ground underfoot. The scratched symbol of a bowl appears in the very first writing, dating to 3100 B.C. Pronounced *nig*, it was used to mean food,[4] and shown combined with an equally cartoonish human head, it became a verb, *ku*, "to eat." Stone was not worried about protecting bevel-rimmed bowls. It was their *images* that were in danger.

In Uruk, the world's first civilization, some were slaves and some were free; more critically, some were freed, perhaps whether they wished it or not, from everyday activities and became full-time specialists. Where the maker of the Brno puppet would have been a person with a gift for ivory carving who nevertheless hunted and cooked, participating in most of the shared activities of their Ice Age band, the scribes of Mesopotamia soon found themselves as insulated from food production and preparation as the ziggurat workers. Their food also came in bevel-rimmed bowls from a central kitchen, while they focused on drawing.

Through two-dimensional imagery, things that had been made into objects were transformed into concepts: a radiant star, a leg with a foot, and waves of water on a stream were drawn in wet clay with a stylus. Then they meant *god* (and also *sky*), *to walk* (and also *to stand*), *water* (and also *seed*, and *son*).[5] Over the following centuries, the scribes abbreviated their scratchings and speeded up. There was so much to record—conquests, imports, directions, a balance of payments. The bowl cartoon would be reduced to four wedge-shaped impressions (the literal meaning of "cunei-form"): three small above one large. Learned and internalized, it had the power to bring not just the idea of food into the mind's eye, but the sound *nig*. Combined with other elements—which were now letters—other words could be spelled. In fact, every word could be written, whether or not it was something you could have drawn a cartoon picture of. To the immortality of an aesthetic sensitivity, projected on Ice Age cave walls, carving had added a means to make a specific utterance last forever.

The neat rows and columns of the little rectangular cuneiform tablets, originally just mud from the floodplain of Mesopotamia, record life in ancient Sumer, Akkad, and Babylon.[6] They are a final, baked-hard testament, sealed in the rubble of the great desert tells, to a way that people once lived, what they thought, and how they thought about it. It was the prospect of things with this special power being lost forever that motivated Professor Stone's protest. The extension of mortal flesh into the immortal inanimate makes us, and has critically made us, human. These written traces of an ancient civilization, like the surviving plays of Shakespeare or Aristophanes, are actually part of the extension of our species' intelligence. Destroying them destroys people, even if not people as usually defined. Stone correctly identified an assault on humanity, an erasing of identity, and a step toward cultural amnesia.

Having been a signatory to a UNESCO-inspired petition sent in the spring of 2001 to the Taliban government of Afghanistan, asking them not to destroy the Buddhas of Bamiyan, my spirits fell when I saw the web link, sent by a colleague, titled "Now you see it, now you don't." It showed the aftermath of the Taliban's explosives work, the hole in the cliff face where the ten-story-high monumental sculptures, which were the tallest standing Buddhas in the world, had once stood. Carved out of a solid rock face in the third century A.D., along with a vast honeycomb of painted monastic cells, passages, and a staircase set into the façade, the images had been deemed *tanweer*, offensive to Islam, by Mullah Mohammed Omar.[7]

As the art historian Hans Belting notes of the various waves of Christian iconoclasm in the history of the West, "Images lend themselves equally to being displayed and venerated and to being desecrated and destroyed"; "injured" images were thought to react like living people, bleeding, or weeping.[8] The neurological underpinning of such a cultural reaction might be that the damaged icon acted as a screen on which the upset emotions of the observer were projected. The outward attribution of inner states to, and generation of inner states by, the exterior image must be connected, as when an erotic image is viewed and physiological changes take place unbidden in the viewer, preparing the reproductive physiology in the absence of an actual human mate. The body is deceived, but not all of the mind, which can remain partly detached in a disjunction of reaction and reality that is destabilizing. This is what another art historian, Aby Warburg, called *Denkraumverlust* (literally "loss of thinking room"), the limiting of imaginative space when the sign for a thing is mistaken, at some level, for the thing itself. Where there had once been an object and a way of signaling the object via a sign, there is now a blur, a confusion. Denkraumverlust is like skeuomorphism on Speed. All societies recognize it (even if they do not name it, or think of it in quite this way), which is why all of them control the way that images are made, displayed, and accessed.

In the Eurasian Upper Paleolithic, for instance, the Venus of Willendorf is one of almost 200 female representations, but comparable male figurines are essentially unknown. There are various kinds of clearly phallic representation, in both cave painting and carving, and there is the clearly male Brno puppet, but nothing really equivalent. One conclusion is inescapable: there were rules and conventions governing what could and could not be shown, what images were approved for projecting in particular social contexts. In later prehistory this trend intensifies. By the time we reach the European Iron Age, with its elaborate material culture of fine pottery and metalwork, it is possible that the majority of objects were gendered (see figure 17)—things without which you could not *be* a woman or a man.

My own research on the Venus phenomenon leads me to believe that, rather than being primarily art—an aestheticized recording of observed reality—the figurines were about creating the earliest social categories. Through the process of formal objectification, the physical production of exchangeable and durable images of women, the actual generalizing category "woman" was brought into play. And just as with

FIGURE 17 Male and female display items from the Iron Age cemetery at Hallstatt, Austria. (Photo: A. Schumacher © NHM Vienna/Bibracte and NHM Vienna.)

fashion magazines today, the projected images became the ideals to which women felt pressured to conform. There is thus a direct line from the external artificialization of femininity in the Ice Age sculptures to the re-embedding of the artificial in the natural that is achieved by cosmetic surgery, from nose jobs and breast implants to leg-lengthening operations. The projective subsystem of the Ice Age was already being used to reflect back and construct an idealized inner reality.

A key element in the Venus figurine phenomenon is not that it is the first art. Recent finds tend to confirm that artistic activities involving cosmetics but probably extending to body painting and potentially including tattooing, extend back much earlier. What is important about the "symbolic revolution" that really gets under way from around 35,000 years ago is that it produced so much that has survived until now. Durability may have been a key consideration in the activities of these Ice Age artists, as if they had begun to grasp that long-term memory needs props.

For the Taliban, the representation of the human form is problematic: philosophically and theologically they have the urge to ban all images; practically they use video to record their activities and generate propaganda. Like every political group before them, they are keen to be influential, and that means memorable. So they, too, cultivate what Gregory the Great called *ars memoria*—the arts of memory.[9] The anonymous author of the *Ad Herennium* (composed circa 86–82 B.C.) treated memory as one of the five parts of rhetoric, and distinguishes between what is naturally remembered (and rememberable) and "the artificial memory," which is aided explicitly by mnemonic devices, frequently potent images.[10] The Taliban aimed to destroy Buddhism in Afghanistan, by wiping its material memory—a technique that dates back to antiquity and the process of *damnatio memoriae:* injunctions on the remembrance of (typically) shamed past emperors, through bans on mentioning their names, through rejection of inheritance claims, and especially "through the destruction and transformation of images of the condemned."[11]

One kind of memory can become another, as the terms of what is being viewed are made to change. The bones of the early Christian child martyr known as Ste-Foy (Santa Fe), beheaded in A.D. 303, occupy a magnificent life-size gold and silver figure, studded with gemstones, in the abbey outside the French town of Conques. But the head of the reliquary seems to be that of a grown man, not a small girl. The chances are that the precious sculpture was originally meant as an image of the Roman emperor, and possibly even Diocletian, under whose edict outlawing Christianity the girl died. We cannot know for certain, just as we cannot know if the bones are really those of the little saint. As Anthony Trollope observes of relics in his short story "Relics of General Chassé": "A stone with which Washington had broken a window when a boy—with which he had done so or had not, for there is little difference; a button that was on a coat of Napoleon's, or on that of one of his lackeys; a bullet said to have been picked up at Waterloo or Bunker's Hill; these, and suchlike things are great treasures. And their most desirable characteristic is the ease with which they are attained. Any bullet or any button does the work. Faith alone is necessary."[12]

Propriety and censorship also govern us outside the world of religion. While there are numerous available photographs of the "last Tasmanian," Truganini, dressed in Victorian style, her skeleton, rightly, has been taken off display, and apparently in tandem, record photographs of what that display case looked like are almost impossible to find (although one exists).[13] The photograph is now as taboo as the skeleton, as it shows it, via the phenomenon of symbolic representation called screen culture. In the previous chapter, I discussed whether the skull used in Hamlet was that of a deceased pianist, or a cast of his skull. The fact is that it was neither—for me, at least: as I watched the production on television, the object was no more than a moving array of pixels. We have grown very used to seeing the world through symbolic representation, and have a tendency to imagine that those who are less used to it are somehow less intelligent.

Intelligence is, in fact, inseparable from the technology that surrounds us. The allegedly objective Intelligence Quotient (IQ) tests, popular through the twentieth century, sought to measure "innate" intelligence (and, in their darkest hours, were a means for labeling different human populations, crudely divided by skin color, as having different cognitive abilities). In fact, as the "Flynn effect" shows, a population's average IQ goes up over time, so it is unlikely to measure anything built-in. Different human populations do have different cognitive abilities, but not necessarily innately so. [They] may appear hardwired because children's brains form themselves concretely in their extra-uterine phase of fetal development, and if that phase coincides with training in abstract signs, such as monitoring the dials that adorn the insides of modern cars, then they will find IQ tests involving the mental rotation of dials easier to solve. Those, on the other hand, who grew up polylingual yet non-literate will be placed at an immediate testing disadvantage.

As Flynn argues, the difference in ability is a shift in cognitive style from the concrete and practical to the abstract and formal. Asked to pick the odd one out from a Labrador, a terrier, and a ferret, we know that while all three are mammals, the first two are interbreedable dogs (*Canis lupus familiaris*) and the latter a mustelid (*Mustela putorius furo)*. But, a generation or so back in my native Norfolk, the Labrador would have been the odd one out: it is a retrieving gun dog, and the other two dig out prey. All three might be grouped with a goshawk to the exclusion of a mouse, in the sensible metacategory "trained animals who assist in the hunt."

The Flynn effect is really an index of homogenization and specific-context deskilling. Deskilling is a complex problem, because our technology increasingly ensures that we need not know anything except how to use the technology. When I caught sight of two paw prints, close together, in soft mud after rain on our vegetable plot (in a community garden or allotment with a serious rabbit problem), I wondered whether my efforts to dig in rabbit-proof fencing to the required depth had been futile. Sarah located the flashlight setting on her mobile phone and I used mine to snap a close-up. Back at home, I was going to check my handbook, *Animals of Britain and Europe*, but Sarah had already punched the keywords "rabbit badger fox cat" into Google and brought up—as it happened—the Northumberland Badger Group website. Only then did we rest secure in the knowledge that it was, as we had thought from the absence of claw prints at the ends of the pads, just a cat.

Perhaps it is unsurprising that I can separate the bones of a rabbit, badger, fox, and cat much more easily than I can identify their tracks: bones are some of the most common finds in the archaeological deposits of the caves I excavate, while I rarely if ever need to distinguish tracks. What was typical of our brief evening trip to the vegetable patch was our practical ignorance, or outsourced knowledge. Humans do not feel the need to carry anywhere near as much knowledge around in our heads as we once did. Technology is changing at perhaps the fastest rate ever, with a continuous pressure to upgrade: I am less shamed by my failure with cat prints than by my clumsy mobile phone touchpad skills.

Near midnight, as I was revising the last chapter of this book, the power failed. There was no moon, and with streetlights extinguished, the town was left so dark that, from above, it would have been hard to notice the presence of technologically competent humans. It was suddenly darker than it would have been in an earlier age of habitual open fires and town lantern carriers—darker perhaps than modern North Korea, where people prepare against the powerless nights. My computer, where, minutes before, I had been entering values into an interactive version of the Drake equation (*N = R* fp Ne fl fi fc L,)* was blank. Yet it was perhaps appropriate that I was forced to write my thoughts about it longhand, by candlelight.

The equation is, according to how it is viewed, very challenging or very stupid. It was constructed in 1961, at the height of the race to put a human on the moon, by the astrophysicist Frank Drake, who wanted to quantify the likelihood of there being other intelligent, communicative life—that is, other technological civilizations—in our galaxy.[14] Until such time as intelligent life is actually located, the equation cannot be solved, and then it will be redundant. Many of its terms essentially remain complete guesses, but it still inspires the Search for Extra-Terrestrial Intelligence (SETI) because of the way it directs attention to the crucial factors. Of particular importance to archaeological thinking (and archaeology's relevance to SETI[15]) are the conditions under which technology might emerge on a planet (the term *fc*) and subsequently endure or fail (*L*).

N is the answer Drake wanted, the number of technological civilizations in the universe that might be able to contact us, and *R* fp Ne fl fi fc L* (in the modern revised form of Drake's original notation) are the factors that have to be multiplied together to arrive at an estimate for it. *R** is the rate of star formation in our Milky Way galaxy. The probability of stars having planets is represented by *fp*, and *Ne* represents the number of habitable environments, planets or their moons; *fl* is the term for the probability of life developing on one of these, and *fi* is the estimated frequency of intelligent life evolving there. A subset of this should be life that has reached the level of technological civilization, *fc*, and thus might produce signs or signals, accidentally or deliberately, detectable from somewhere else in the universe. *L* is the duration of such potential detectability.

Various terms of the equation, for which Drake originally had little or no actual data, are now coming into focus. Recent advances in space telescopes have identified other planetary systems, and although it is not yet possible to see Earth-sized objects, statistical analysis indicates they will be found, and also that up to half of all stars may have orbiting planets. But estimating the likelihood of life on them is far harder. Estimates of which are potentially habitable, *Ne*, are interdependent with estimates of *fl*, the probability of life evolving. This is because habitability depends on what form of life we are talking about. Even if we solve this, there follow the questions of intelligent life, of technological emergence, and of the visibility of advanced civilization.

Considering *L*, it is thought that our species had probably become detectable by around A.D. 1950, when radiation associated with radar and TV broadcasting began to leak out into space. As that is sixty years ago, we now sit in the middle of a bubble of detectability some 120 light years across. On present estimates, there are about 100 billion (100,000,000,000) stars in the Milky Way galaxy, the vast majority farther away than this, so our civilization might collapse a million years before any signal from us is received elsewhere.

It is in the nature of technology to allow such temporal disjunction. A prerecorded voice of someone I have never met (who could, biologically speaking, be dead) abruptly emanates from the telephone answering machine. The lights all come back on as power is restored. Only when electric power is absent do we realize how much we take it for granted, and we do the same about life. On this planet, it seems natural. Indeed, that is our word for it. But is it commonplace in its cosmic appearance, or fitful, erratic, perhaps uniquely rare? How likely is technology, and how likely is it to fail? *L* has to capture the probability factor of a blackout with the effects being more than just local and more than just temporary—an interruption in entailed technology that would plunge us as a global species into the kind of regional isolation that the Tasmanian Aborigines found themselves enduring for more than 10,000 years.

At the outset of this investigation, I identified three systems as governing everything we know. System 1 is inanimate and natural, System 2 is natural and biological, and System 3 is artificial and technological. In terms of the Drake equation, it is the first system we now have the best grasp of, being able to say that locations with the correct physico-chemical conditions for life are likely to abound across the universe. But *fl*—the likelihood of life—is much less easy to assess.

The exclusive reign of System 1 ended in what is known as the "late heavy bombardment," a formation event that cratered our moon between 4.1 and 3.8 billion years ago and that must have pummeled the earth equally hard.[16] Yet life appears to have emerged on earth at or around 3.85 billion years ago, just as soon as the heat and impact conditions had been relieved by the smallest fraction.[17] Perhaps the space debris itself brought the basic building blocks of life with it, as the astronomer Fred Hoyle speculated, so that the universe somehow seeds itself with life (Hoyle's concept of "panspermia").[18] The chance of a single DNA molecule randomly assembling itself somewhere in the universe is estimated as around 10^{130} (while the number of atoms in the observable universe of over 80 billion galaxies is only 10^{80}). If that would appear to make life very unlikely, simply being ignorant (as we remain at present) of the stages of early synthesis does not mean it was not a nonrandom, highly likely, perhaps inevitable process, because we now know that life on earth emerged just as soon as it was physically possible for it to do so: System 2 may be an inherent, universe-wide condition.[19] If that is so, then it is also significant that, as Lewis Dartnell of University College London has recently pointed out, the survival conditions for life on earth demonstrably overlap with known conditions elsewhere on planets and planetary moons in our own solar system.[20] That is, the number of extraterrestrial environments where organisms of a type that already exist on our planet could survive is now known to be several, notably on Mars and the moons of Jupiter.

In fact, we already know that life *has* existed elsewhere within the solar system; it did, twice in 1969, twice in 1971, and twice in 1972, when the most complex artifact ever assembled—the Saturn V rocket—landed astronauts on the moon. The archaeological evidence is still there: Alan Shepard's golf balls, from his famous Apollo 14 mission driving practice, lie on the moon's surface and betray the telltale structural patterning of a System 3 object. There, on the moon, wearing a PLSS (Portable Life Support System), human beings have eaten, drunk, laughed, jumped around, and declared the significance of being a smart biped ("one small step . . .").

As technology changes, our exposure to observation from beyond our own solar system alters. The current shift from high-wattage broadcast television to low-energy fiber-optic communication already makes us less visible than we were. I am less visible than I was too, as I have made the choice to switch the halogens back off and continue by candlelight, drafting for now in pen. My bronze candleholders, though not original, are of an eighteenth-century style that would have been familiar to anyone aboard the *Beagle*. A present from my late godmother, they remind me of her and prompt me to notice all over again that technology embeds our sentiments and our memories far more deeply than we normally admit.

The act of lighting these candles has brought Rosalind to mind as keenly as her photograph did earlier this evening. The snapshot showed my younger daughter as a baby, scrutinizing Rosalind. She does not now remember my godmother directly, but photographic technology is a time-factored mirror. It reflects a past as it existed before a time that Josephine's organic brain can distinctly remember, although she was there. The technology thus extends her intelligence in time, and aids the formulation of memory. This is what archaeology as a discipline does in deeper time when it uncovers happenings from time out of mind.

When the child whose skeleton we found in Y-Pot was alive, the first metalwork had come into being, and the pace of communication was already rapidly increasing through the invention of the wheel and the domestication of the horse. But there was no possibility of communication with other worlds except those imagined ones that were the dwelling places of the gods. Although the Bronze Age is now 3,000 years ago, the shape of the modern world was already becoming clear. In a sense, the critical moment had passed over 2 million years previously, when we first became technological and technology—in particular in the form of the baby-carrying sling and of cooking—allowed for the production of underdeveloped, small-headed babies through narrow pelvic canals, and their extraordinary extra-uterine development in a protected, high-protein environment, where their brains could grow massively after birth.

My estimate of the Drake equation value is unusual, perhaps, in that I am prepared to ascribe a high probability to *fl*, *fi*, and also *L*, but a very low one to *fc*. Whatever the actual numbers, we are increasingly able to observe other solar systems, we know life began early on our planet, and we know life is robust enough to withstand existence elsewhere. I think there is a good chance that extraterrestrial life will be detected before I die, probably as a result of observing a distant planet on which photosynthesis appears to be happening. And as the second system, that of biology, should conform to Darwinian rules wherever in the universe it is found, there will be competition among organisms, and a premium on the emergence of intelligence. Intelligent life, in turn, is likely to modify its environment in the way that birds do in building nests and spiders weaving webs. It is also not unlikely, in my view, that innate, genetically hardwired abilities to manipulate materials might give way to yet more intelligent behavioral routines transmitted by learning rather than instinct. In short, I would not be surprised if plenty of cosmic life existed at an intelligence level similar to that of dolphins and chimpanzees, neither of which can be imagined to have the potential to become communicative in interstellar space. If half the 100 billion stars have one habitable planet on which life can emerge, my guess is that it naturally will, and will naturally evolve intelligence; but the chances of a technological, communicating civilization seem very small.

I have argued that a unique series of conjunctions was required to bring about the biologically paradoxical evolution of a large-brained technology-using biped. I have suggested that technology appeared initially before significant increases in intelligence, and actually allowed its emergence. There is no reason to believe that this peculiar sequence of events was in any way inevitable, and the time that elapsed from the first beginnings of upright walking some 5 million years ago, to the appearance of the first chipped stone tool, saw no significant advance in mental capacity, and species such as the chimpanzee stayed only as smart as our last common ancestor.

While I think that the chances of intelligent life developing technological civilization, and thus a new form of intelligence, which we can term "technological intelligence," and which must underpin the sort of cosmic communication Drake speculated about, are extraordinarily small, I am optimistic in believing that despite all the environmental

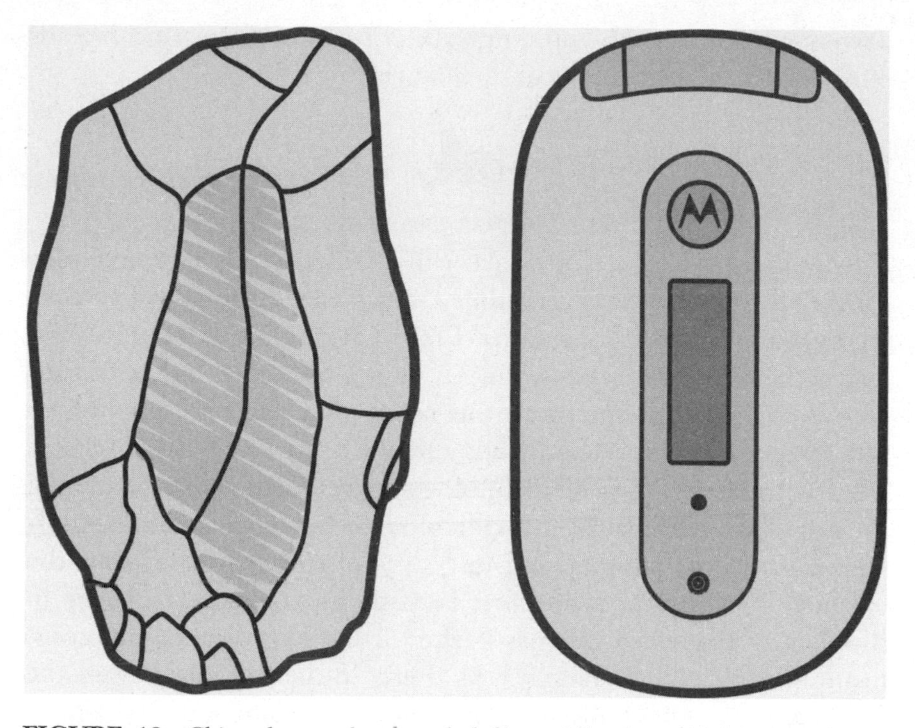

FIGURE 18 Chipped stone hand tool (left); mobile phone (right). (Graphic © Frankland.)

challenges we now face, such civilization and such intelligence will prove extraordinarily durable.

The whole debate about whether computers can think, or will be able to think, or have consciousness, is stagnant. It depends on an outmoded understanding in which only two systems exist, the inanimate and the animate. In fact, System 1 (inanimate things and things that were once animate but are now dead) and System 2 (living things, including beings conscious enough to understand intentionality, to signal, communicate, and deceive) are complemented by System 3 (artificial things, including books and computers, which are potentially immortal, containing intelligent information in nonbiological forms). A deer and a human baby are animate and alive, while a pebble and a mobile phone are inanimate—not dead, but non-alive. Yet the non-aliveness of a pebble and the non-aliveness of a mobile phone (for poetic reasons, let us say it is a Motorola PEBL; see figure 18) are very different. As a bipedal, hands-free ape we can use both of them, of course. The pebble can extend a zone of protection in an outward radius from our bodies by being thrown, or by a throw being threatened; and it can, rather crudely, attract another person's attention, make a splash, smash a window, hit them. The other interfaces with our brains and extends our intellectual power in far more nuanced ways.

Technology is magic. "Now we can wirelessly connect and play games with each other!" exclaimed my teenage daughters as they started running a program called Pictochat on their Nintendo DSs: just as they had done when younger, they altered each other's scribbles to make surprise pictures. But where they had once used physical scraps of paper, passed back and forth, now it was on screen. That it took this level of technology to get two siblings living in the same house to play seems ridiculous... except that the technology is the whole point. Good or bad is often hard to say, but the balance of enchantment and fear is easily upset, as on the beach in Blackman's Bay when mirrors flashed from fantastical to ominous, turning in Aboriginal hands. Was there, in their minds, before the violence broke out, some premonition of the coming global age of screen culture, of a reduction to surfaces, to repetition—an endless sale of identical images?

The *Beagle*, an expedition vessel adapted from a fighting brig, had plenty of guns on board, including cannon, and used them at times to intimidate natives. Darwin deplored what he saw as the inferior, primitive intelligence of the Fuegians, writing in his journal, "It was most ludicrous to watch through a glass the Indians, as often as the shot struck the water, take up stones, and, as a bold defiance, throw them towards the ship, though about a mile and a half distant!"

The naturalist interpreted this as innate stupidity—a race-based cognitive deficiency. Having since challenged our own perspectives and attempted to see things differently, perhaps we can appreciate how the *Beagle* must have challenged Fuegian perceptions. It had to do with scale and expectation: compared to the single-element hollowed-out log canoes that were all they knew, the *Beagle* was gargantuan. But that would not have been apparent. Being clearly a watercraft, it must at first have been perceived as much closer to shore than it really was. Sight is in the mind as well as the eye. By throwing stones, the Fuegians were range-finding, scanning what happened and revising their mental map until the *Beagle* suddenly popped into its true size;

not a funny sort of canoe, but a ninety-foot-long brig, with composite masts that rose over the ocean higher than most Patagonian trees, noisily spitting fire.

As we saw in chapter 1, for all their difficulty in perceiving an utterly alien object, the Fuegians had very acute vision compared to the crew of the *Beagle*. While some people claim that the human eye is so perfect that it must have been made by an intelligent designer, vast numbers in the modern world wear glasses, many suffer from cataracts, and optometrists agree that, even at its healthiest, the "design" of the human eye is very poor. That is why we design spectacles, cameras, telescopes, microscopes, to correct and enhance our vision. Optical technology not only compensates for our natural deficiencies, but also allows us to stretch beyond the natural. With optometry and lens manufacture, we no longer need keen sight in the way our hunter-gatherer ancestors did.

Human eyes, in the modern world, have degenerated—but our vision is better than ever. Supreme accuracy with a bow and arrow under a clear hunter's moon is not the guarantee of dinner it once was. And bringing the stag to the feast is no longer the best way to secure a fit partner with whom to produce children who will inherit their parents' visual acuity. With a well-stocked fridge at home and spectacles firmly in place, even the desperately shortsighted can go forth, fall in love, and multiply. Their children may have even worse eyesight than their parents, but can easily avoid deer hunting and find success elsewhere.

Human vision is no longer just biologically supported. We have a combination of visual sensors, some organic, gene-based, and biological, others linked to ground glass, hydrated silicone, and fiber optics. The food of technologically developed humans stays fresher than the food of hunter-gatherers, and we can use a greatly enhanced power of vision to identify the pathogens that cause it to become toxic.

Humans, continuing to evolve biologically and in some senses degenerating and weakening, are growing dramatically stronger. What removes the natural pressure to be born with sharp eyes is, most directly, the technology of optics. And optics allows the identification not just of germs at the micro level, but of meteorites at a macro level. Developed through satellites and missile tracking systems, bar scanners and retinal ID analysis, optics is a key element in many of our greatest successes. It protects us against the external environment and provides a competitive edge over less technologically advanced groups. So those with access to telescopic sights on guns prevail over those without, and those with remote guidance systems prevail over those with mere telescopic sights. Overall, technology produces the environments within which fitness is ultimately judged, regardless of nature. An insect-catching bird may have natural visual acuity hundreds of times finer than ours, but we can track it, catch it, tag it, trap it, or kill it at will. Then we can study its eye and use the lessons we learn to design new things.

Our great failure to be well-adapted to any particular environment, which forced us to survive by building the objects we require to stay alive, turns out to be our great strength. We have expanded all over the globe precisely because we are *not* adapted to survive in any particular environment—at least, not survive unaided. But, aided by the things we can make, from spears to guns, from fur hats to Jet Skis, from skyscrapers to satellites, we can succeed everywhere and anywhere. All we have to do is set our minds to it, and build the supports that our fragile bodies need.

Darwin knew he had gone wrong when his instincts contradicted his own theory. By his own logic he should have praised the feckless breeders of the Victorian underclass. The algebra was obvious: in terms of natural selection, the sacrifice of quantity for quality was the better strategy. The undernourished children of large families, brutalized by cramped living quarters, emotionally scarred by the commonplace of sibling death, ill educated and often delinquent, were nevertheless outcompeting the healthier and better-educated offspring of smaller families. By Darwin's own theory, the latter were not then actually healthier. The logic of survival of the fittest meant that the evolutionary successes were those great extended families of children; yet he saw them as a threat to humanity.

Darwin understood the tension between an apparent biological imperative and alternative, humanly defined measures of success when he considered whether or not to marry, jotting his thoughts on the subject in one of his notebooks. There, in his now-famous two-column pro and con list, he wrote "less money for books &c—if many children forced to gain one's bread." He decried the unregulated breeding of the urban poor and the contrasting restraint shown by the professional and intellectual classes, who had to choose between having lots of children and being able to afford the books with which to educate them. Darwin never squared the circle here: by his own logic, the urban poor were fitter. They passed their qualities in greater amounts to the next generation, and so were better suited to their environment.

Darwin seems to have been unable to accept this. Certainly his thinking on the subject was not particularly lucid—he was wealthy enough to have large numbers of children *and* books, and seems personally to have avoided the reproductive dilemma of the aspirant middle classes. Ironically, the resources for this luxurious strategy were based in large part on a financially advantageous cross-cousin marriage with Emma Wedgwood, of the Wedgwood pottery family. The company founder, Josiah Wedgwood I, and his wife, Sarah Wedgwood, were maternal grandparents to both Charles and Emma. The biological result seems to have been a marked genetic infertility. The Darwins had ten children; seven survived to adulthood and six married. Three of these, despite being physically normal, wanting children, and living long, healthy lives, remained childless. Molecular research shows that this type of inherited infertility can be attributed to inbreeding. It is particularly tragic to note the childlessness of Major Leonard Darwin, who lived to ninety-three and married twice, and whose book on eugenics stressed the importance of "right breeding" to an Edwardian audience.

It is clear that Charles Darwin believed that the future success of humanity would be assured by quality not quantity, and by the power of the mind rather than by muscle power and weight of numbers. It was a vision that tacitly accepted that natural selection could not, and should not, apply to people. In seeking to influence reproductive behavior and promote the proper education of children, he was giving the lie to the pile-'em-high, sell-'em-cheap algorithm that could provide the greatest return in mere biological terms.

Darwin was paying attention to the value of human culture, and to the mechanisms by which we had managed to raise ourselves up from a bestial state. If he was too busy working out the theory of natural selection among other animals to pay attention to how it was ultimately subverted by the material world of artifacts, we cannot blame him. Pursuing such a line would have weakened his argument that we were descended from the apes and the revolutionary insight that we shared a broader ancestry with all other living things.

The human brain is a liability—like the peacock's tail, it is a potentially unnecessary, expensive-to-maintain feature that is energy-inefficient and creates physical vulnerability. This is particularly so during childbirth and in early childhood. Were the balloon-like heads and tiny bodies of newborns not so familiar to us, we would be staggered by their appearance. Our offspring can hardly move due to a huge cranium balanced upon a preposterously weak and spindly spine (unlike that of any other mammal). If a small child's head is moved involuntarily (either by accident or through the kind of aggression seen in "shaken baby syndrome"), death may quite easily result.

In basic terms, human brains are three to four times larger than those of our primate cousins, and therefore cannot be a simple competitive response to them. A comparison with the big cats, another group of animals that evolved on the African savannah, helps demonstrate this. The cheetah is fastest, with an estimated top speed of 65 mph; the lion is next, reaching 50 mph on short bursts, and outrunning the cheetah over longer distances; the leopard has a top speed of around 45 mph, having a skeleton that allows it to clamber into trees, which cheetahs cannot. Cheetahs can outrun antelope in the open, whereas leopards tend to ambush them from close range; lions use a combination of stealth, stamina, and pack-hunting. These three big cats share a common ancestor, evolved in competition with one another in the same environment, and have similar attributes. This is true for their prey species, too: none of them could afford to be radically slower than the others, and they fall into the same range as the animals that hunt them (pronghorn antelope, 60 mph; springbok, 55 mph; Thompson's gazelle, 50 mph).

If the cheetah's top speed was 200 mph (four times the lion's), we would be very puzzled. Why would it need to keep getting faster if it was already the fastest? That is the sort of difference in magnitude that exists between our brain size and that of chimpanzees and gorillas, with whom we shared a common evolutionary ancestry in Africa in the period before 6 million years ago. And we know that (long after this common ancestor) the australopithecines, from whom our genus eventually emerged, retained brains in the same, standard, chimp- and gorilla-size range.

The super-size brains that eventually emerged cannot simply be a response to competition: once protohumans had outstripped rival apes in intelligence, pressure for further enhancement would have vanished. One possibility is that we were driven by a need to become ever smarter hunters by super-smart prey. But that argument does not stack up on its own: chimpanzees have little trouble outsmarting smaller primates, such as spider monkeys. They have significant social intelligence that allows a fearsome level of organized hunting, using sticks and stones to augment their ripping teeth and powerful musculature. Perhaps what drove human brain size upward was an internal quirk. Perhaps it was competition between ourselves, between our precise genetic line and a whole series of also-rans.

One—perhaps bleak—speculation is that upward-spiraling intelligence was driven by technological warfare. The evolutionary arms race that brought about our species' massive increase in brain capacity was a real arms race, as different species and groups battled to develop and control ever more lethal technologies in a process that favored physically weak, but increasingly clever, individuals. Our last main competitors, the big-brained Neanderthals, were finally annihilated only around 30,000 years ago. Perhaps that is why, since the Neanderthals were exterminated, human brains have been gently shrinking. Our brains are becoming smaller but not more efficient, but neither are we, as a whole, less smart. In fact, it is the reverse: the emergence of a powerful symbolic culture, leading from art to writing and—eventually—the digital age, increasingly allows our intelligence to be farmed out.

Of the seven great apes, only one gets active after sundown, observes the behavior of its fellows via TV, fixes a late-night snack from the fridge, and browses the web. These are the things that make us what we are in a day-to-day way, and that have made us what we are in terms of our physical deficits and intellectual oddity. We are formed not by raw nature, but by the continually emerging world of technology. It seems that artificial intelligence *is* human intelligence. The idea of humans versus technology is wrong. The vision of computers becoming far more intelligent than "us" is misleading because they are us too. We have to face our destiny as the first nonbiologically evolved species. Technology is at least as critical to our identity as our soft tissues.

Vilém Flusser, the Czech philosopher, is known for his provocative investigations of how the physical form of external things and the shape of our thoughts reverberate with one another. In his essay "Shelters, screens and tents," he wrote that "a screen wall...is a wind wall. A solid wall...no matter how many windows and doors it possesses, is a rock wall. Thus a house, like the cave from which it derives, is a dark secret...and a tent, like a nest in a tree, of which it is a descendant, is a place where people assemble and disperse, a calming of the wind."[21] The development of screen culture now insulates us even from the winds of time and chance. Although the Taliban destroyed the Bamiyan Buddhas, they have, for now, made them more timeless and enduring than they ever were. Added to the massive archive of photographs of the mural paintings curated by the University of Vienna Art History Department, there is now a project to create a virtual museum set in a virtual Bamiyan.[22]

CONCLUSION

ONE WORLD, AGAIN

Sadly we don't even have a good theory about technology.
—Kevin Kelly, *The Technium and the 7th Kingdom of Life*[1]

HUMAN CULTURE IS vastly complex, and I have threaded a particular course through it, touching only indirectly on the importance of language and the creation of social institutions. Whole books have been written that attempt to define what culture is, but a very useful account, especially when viewing humans in a really long-term perspective, is that by the anthropologist Ernest Gellner. Gellner pointed out that the human species displays "an unbelievable degree of behavioural plasticity or volatility"[2] but that this genetic "under-determination" is itself genetically determined. In short, our genes give us the potential to be all sorts of things but leave us incomplete.

Paradoxically, Gellner argued, this very behavioral plasticity deprives us of so much innate, instinctual know-how that, to survive, we have to learn how to be after we are born. This process of childhood enculturation physically molds the infant brain as it soaks in the human world from the safety of its sling. No surprise that, once established, what we have learned of our own culture is almost fascistically adhered to as "the way." So humans, taken en masse, are incredibly diverse while, viewed within their cultures, they are incredibly conformist. The behavioral plasticity of childhood gives way to the normative values of adulthood, which by then have been internalized as "natural."[3] In some ways, humans, like hermit crabs, must find their niche; critically unlike hermit crabs, this niche is not limited by the size of vacant mollusk shells.[4]

Chekhov, on his visit to Sakhalin Island, reflected on the disjunction between the lifeways and perceptions of the local Gilyak hunter-gatherers and the Russian settlers: "How difficult it is for them to understand our way of life may be seen from the fact that they have not yet completely understood the purpose of roads. Even where roads have already been built, they continue to travel over the taiga. Often you will see them, their families and dogs laboriously making their way in single file through the marshes beside the road."[5] Intelligent as he was, Chekhov was unable to see that his "natural" perception of a road was not the only way of seeing such a technological structure. For him, it was to aid movement from A to B, and the Gilyak failed the IQ test. Viewed in Gellner's cultural terms, what was at stake was a different set of ingrained categories of thought. The Gilyak did not spend their time awaiting arrival somewhere else; they were always somewhere. For them, the road was nowhere. If movement was process, then the road was an obstacle to perception, a zone where animal tracks could not be traced or plants identified; not wishing to be insulated from the sound and feel of the actual earth, the Gilyak shadowed it.

Further shifts in road perception have occurred since Chekhov. Traveling on the interstate guided by satellite navigation, we follow an on-screen line and have no direct contact with our surroundings at all. Modern technology here achieves what the playwright Max Frisch presciently called "the knack of so arranging the world that it no longer needs to be experienced." Nowadays, those of us living in the developed world hardly need to use our bodies. Without breaking a sweat, we manage the basics that often took our forebears every waking hour, and accomplish tasks that they would have considered magical—flying between continents, communicating across thousands of miles, preserving the movement of the past not in an inherited dance, but in Blu-ray digital. But something else has happened too: satellite navigation involves GPS and the roll-out of a single technological standard available, potentially, to all cultures. The Internet holds the promise of all the world's individuals socializing again around the same campfire.

On the face of it, it is ironic that the ultimate technological driver of the communications revolution has been military. Weapons technology—from those first showers of slingshot "hand axes" to the remote-controlled drones that can be guided by an operative in front of a screen in an office on one continent and targeted at an insurgent training camp on the other side of the world—has finally delivered the hardware that allows us, potentially at least, to live as one interconnected species. From the original single template of the chipped stone tools, used 2.6 million years ago at Gona, humans evolved and diversified across every environment on the planet. Now we are conforming them all to the same technological standards: one world, again.

In *The Brisbane Courier* for Tuesday, February 24, 1891, the following report appeared: "News has been received from Fiji that the A.U.S.N. Company's steamer *Truganini* was totally wrecked on the island of Aneityum on the 11th of February during a hurricane. The crew were all saved but the cargo was lost. The *Truganini* was a three-masted iron screw steamer, 120ft. long, 20ft. beam, and 9ft. depth of hold. Her engines were compound, two cylinders, one of 14in. and the other 25in., both having a stroke of 18in."[6] The tough little coal-fired ship designed to exploit tin and gold on the hard-to-access far west coast of Tasmania had been built in 1876, the year of the Aboriginal Truganini's death.[7] The name was less *In Memoriam* than *In Dulci Jubilo*, perhaps with an edge of mockery. Certainly there had been sweet jubilation in the press when the native race, deemed too mentally retarded to have had either clothing, boatbuilding, or fire-making skills, appeared finally extinct.

A woman who had been scrutinized naked and permanently recorded in the form of the little limestone figurine found at Willendorf was now joined in the long history of objectification by a woman whose skeleton was sequestered for a grotesque museum display and whose name, which had meant "toolmaker," was appropriated for an item of industrial technology used to transport metal to make tools.

The first anatomically modern humans in Europe and the first in what would become Tasmania arrived almost simultaneously, between 40,000 and 35,000 years ago, carrying much of the same cultural patrimony: stone tool use, baby slings, fire-making and cooking, spears and spearthrowers, symbolic art. That is no surprise, because they had evolved as the same species, developing out of the African *Homo heidelbergensis* population sometime after 200,000 years ago. But by the time the first

Europeans reached Tasmania the technological trajectories via which each group had reached the present moment could not have been more different, in outward appearance or inward logic.

At the start of this book, I suggested that there were three systems governing all the patterning we see around us; those of a religious bent may prefer to argue for just one—divine power—and physicists may essentially do the same, hoping ultimately to unify all forces, and so all causes, in a single algebra. Darwinian biologists prefer to view life, at least, as special, and separate from nonlife, and so they recognize two systems, that of inanimate forces, and the competitive, evolutionary system of living things. But they also have a tendency toward reduction and would like, if they could, to provide a satisfactory theory of how human culture could be conformed to Darwinian laws. As technology is a visible, and apparently measurable, form of culture, the idea that technology might be made up of networks shaped by competition, adaptation, mutation, and reproduction, not simply analogous to Darwinian natural selection but identical with it, appeals to them. My argument, that the realm of material artifacts brings a new kind of patterning and new kinds of variation into the world, is riskier perhaps. Certainly it is less parsimonious. [Kevin] Kelly rightly remarked, "we don't even have a good theory about technology" (and this book can only be a minor contribution to building one.)

Technology is the know-how to shape things to achieve ends that appear impossible. Typically, it does this by insulating us from the effects of a particular, uncomfortable, or impossible-to-survive-in environment, providing a portable "pocket environment" that meets our needs: a fur coat for a cold day, canned food for a polar expedition, a giant rocket-propelled tin can and a space suit to take us around to the dark side of the moon. But if those are needs, then our needs are not the same as those of other species. And even what we consider to be basic, biological needs seem to have arisen in tandem with our emergent technology. The intelligence that makes us inventive was enabled by inventions: the baby sling, the stone blade, the cook's hearth. These are not the same as inanimate, natural things. They are artificial and form the nonbiological aspect of the artificial ape.

The claim that technology has taken the leading role in human evolution is an apparent paradox, as what is most remarkable in our emergence is our development of an incredibly powerful intelligence—which we use to invent and use technology. The argument is made on the basis of a very fragmentary survival of objects from the ancient past.

But the traces of past biology are fragmentary too, and when discussing the increase in brain size, although we have many skulls, we essentially have, from the entire period of the earliest human emergence, just three examples of the kinds of pelvises through which the skulls would have had to pass. We have chipped stone tools from 2.6 million years ago, but organics, notably artifacts like carrying devices, turn up only in the most exceptional circumstances. If it wasn't for finds like Ötzi, preserved in ice complete with his bentwood rucksack, or the Bronze Age mine at Hallstatt, with its dry, salt-permeated conditions where leather-and-wood carry sacks survive complete, we would know very little about prehistoric carrying technology. Yet we would have to infer its existence.

Looking at the pattern of increasing brain size among our various ancestors and their hominin competitors (see figure 19), we have to ask the question: At what point did the baby-carrying sling first appear? Like a pin-the-tail-on-the-donkey game, we could put our marker almost anywhere. Dean Falk, in her consideration of the problem, at first considers that *Homo erectus* mothers must have used them, and then imagines that perhaps they did not.[8] In the absence of any preconceptions, however, there is just one place on the graph where the invention, so placed, would act in the revolutionary way I have imagined—in the period of rapid brain-size increase just after 2 million years ago.

Let us review the trajectory: From 5 or 6 million years ago, one and then more species of hardy little apes tried out their hind legs when they were not nesting in trees or swinging through branches. By 3.4 million years ago, footprint evidence shows that australopithecines of the Lucy type could walk pretty well over distance, and by 1.8 million years ago, fully upright walking had emerged. What happened in between was the appearance of technology, at 2.6 million years ago, leading to the emergence of a genus—ours—that began to be able to spread out of Africa.

After bipedalism emerged, along with a new sideways-oriented birth mechanism, childbirth seems to have stayed essentially the same in early bipeds, including our genus, judging from the laterally organized birth canal in the Gona pelvis (if it can, indeed, be attributed to *Homo erectus*); that is, for the present, the change to a modern style of rotational birthing does not occur before around 0.5 million years ago.[9]

So, for the first 2 or 3 million years, any head-size increase is locked down. Brain size increases massively just after 2 million years ago, at a time when we know that technology, System 3, is already up and

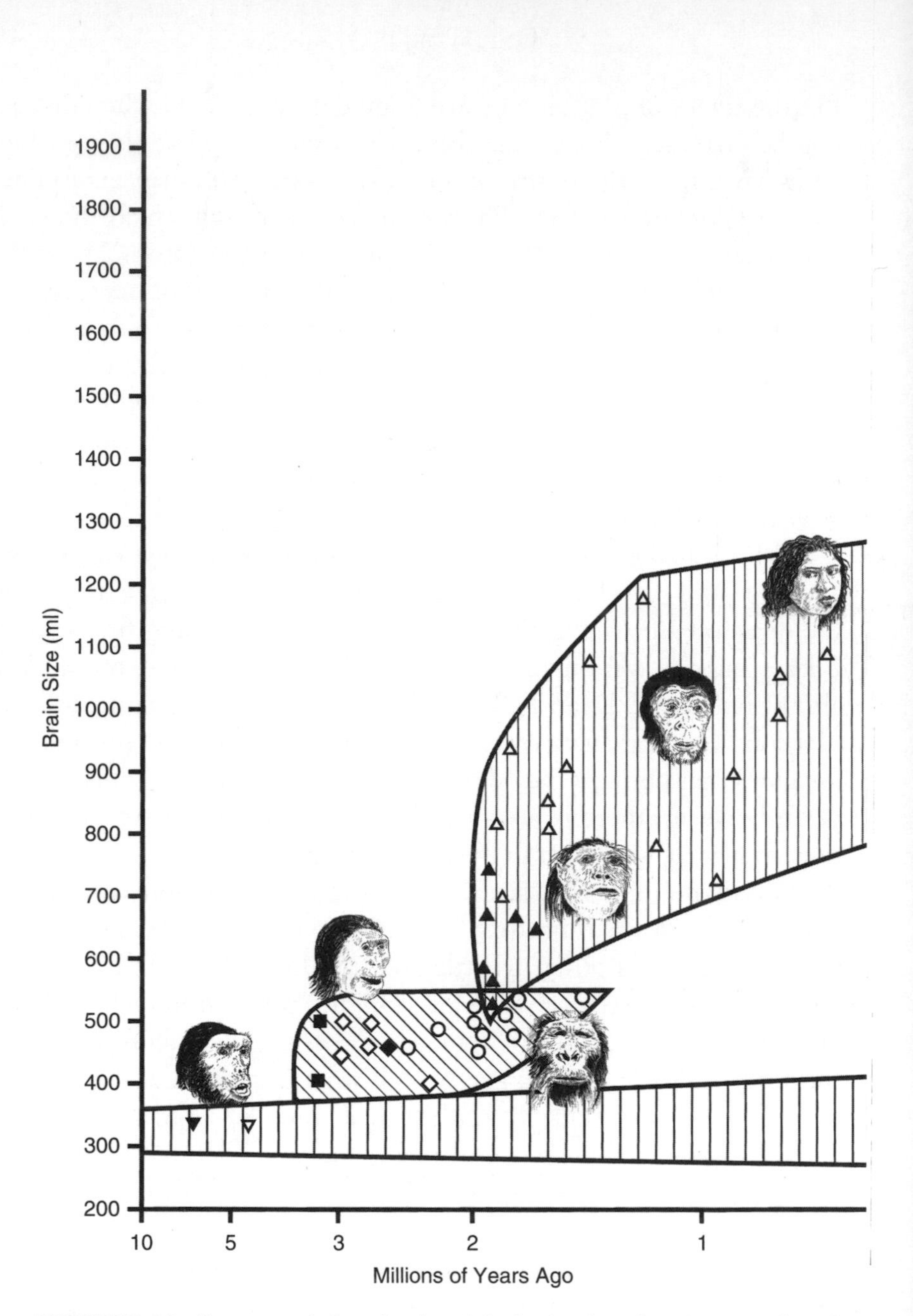

FIGURE 19 Interpreted data for hominin brain-size changes over time (see figure 5 for symbol key). Wide vertical shading = stable brain-size trajectory for inferred chimpanzee ancestors; diagonal shading = australopithecines and paranthropines; narrow vertical shading = rapid brain-size increase in early genus *Homo*.

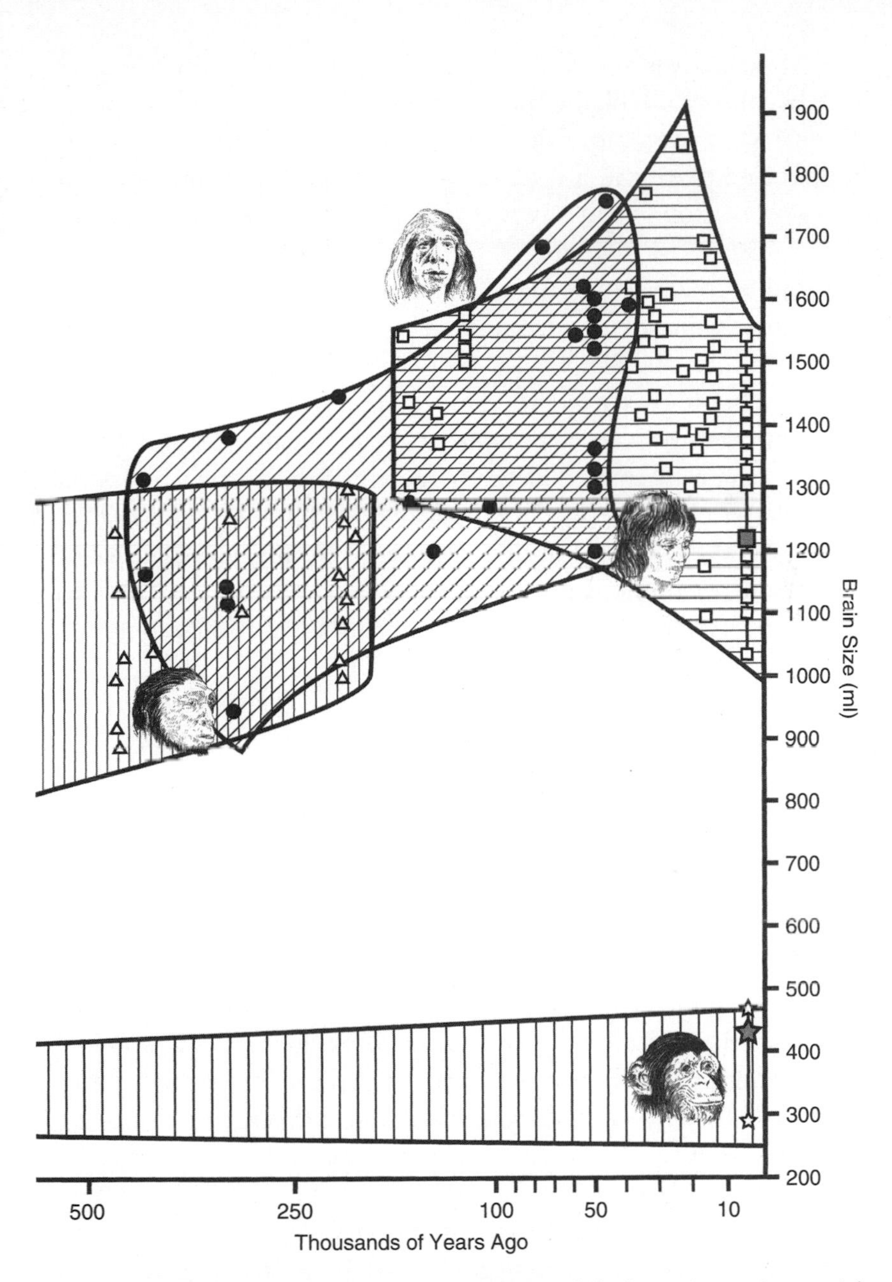

Narrow vertical shading = later *Homo erectus*; diagonal shading = increasing cranial capacity of *Homo heidelbergensis* and Neanderthals; horizontal shading = increase and subsequent average decrease in *H. sapiens* brain size. (Graphic © Frankland/ Taylor.)

running. My conclusion is that the rapid expansion in brain size after this time becomes possible because hominin females are by this point able to deal with ever more developmentally retarded young, and I would therefore place the invention of the sling as coterminous with the emergence of our genus. It is, I believe, the thing that makes us human.

Half a million years ago, changes to the pelvis allowed brain capacity to reach regularly into the modern range, but our skeletons remain very robust for another 300,000 or 400,000 years. From around 100,000 years ago, technology started to remove significant strain from the skeleton, a process that accelerated after 30,000 years ago and accelerated faster again after 10,000 years ago. In this shortish period, a cascade of material innovations took place, following hard and fast on each other. By 1987, when the first version of Honda's bipedal robot Asimo was built, technology itself could walk bipedally. In 2010, Asimo can run, both feet off the ground at once during forward movement.

A century and a half after Darwin, with evolution a provable fact and its genetic underpinning in the recombinations of DNA laid open for future manipulation, many see a turning point. Geneticists like Steve Jones say that natural selection is potentially over—the future can increasingly be designed. Technological visionaries like Ray Kurzweil argue that the age of the machine is upon us, and that humans will fade into obscurity as technology reaches a point where it can innovate itself, producing ever more complex forms of artificial intelligence. My argument in this book is that, scary or not, none of this is new. Not only have we invented all technology, from the stone tools to the wheeled wagon, from spectacles to genetic engineering, but that technology, within a framework of some 2 to 3 million years, has, physically and mentally, made us. We long ago began adapting our minds and bodies to a hidden agenda. The result is a new, symbiont form of life—one that breaks the old rules.

The philosopher Simon Blackburn writes, "Whatever may have been the case among our ancestors back in the Pleistocene and beyond, today, among adult human societies at work, artifice lies in our natures."[10] If humans have unwittingly engineered themselves, the implications are wide ranging. The claims of Ray Kurzweil (that we are approaching a critical moment when biology will be overtaken by artificial constructs) and Joel Garreau ("For hundreds of thousands of years our technologies have been aimed outward, at modifying our environment...now we've got a suite of technologies that are aimed inward, at modifying our minds, metabolisms, personalities and children"[11]), while making some sense, lack a critical historical—and prehistoric—perspective.

They are right to draw attention to the potential impacts of new technologies—impacts they welcome. But they are wrong to claim that anything fundamentally new is happening. Constructing an artificial environment and forming ourselves in artificial images is what humans have been doing for at least 2 million years. Perhaps the fundamental difference made by the past few centuries of intellectual endeavor is that, via the discoveries of archaeology and paleoanthropology, we can now review the many stages by which we became human. Yet instead of finally sinking our desire for gods, the new vision of evolution and the tools of genetic engineering it has produced suggest a future so uncertain, challenging, and unfamiliar that many are being driven back to their spiritual comfort zones. The return to blind faith is so abrupt and violent that it threatens the foundations of our enlightened scientific knowledge of the world and our place within it—classical foundations that were reset with difficulty after the superstitious upheavals of the Middle Ages.

We can never escape the bio-technological nexus and get "back to nature," because we have never lived in nature. But there is something potentially wrong with our technology: it is dangerously entailed. The distance between cause and effect has become so great that by the time we perceive our potential maladaptation to environment, it is too late.

Darwin was right to try to resist the terminology "survival of the fittest"; biological evolution is littered with the extinct fossils of apparently perfectly adapted creatures. And the same is true for the relics of culture history: the great civilizations of Egypt, the Indus Valley, and the Maya. Entailed culture always demands something more, real or imagined. The Maya, in cities like Copán, built temples on the flat, in the fertile valleys, and offered their gods powerful hunted beasts from the surrounding rainforest—the finest specimens of jaguar they could find. Only minor changes in climate, coupled with an apparently successfully rising population, threw them into crisis. As shortages led to social unrest, the temple-building program was accelerated in order to pacify the gods. In reality, it simply removed more productive land and labor from the system at the time when it could least afford it. But the whole infrastructure was based on belief in averting chaos by conformity to the will of the deities, and the procedures of technological fabrication tended to metastasize into the routines of ritual, as we saw in the Upper Paleolithic symbolic revolution.

"Sustainability" is a word that is meaningless without a specified timescale. Unless we know how many months, decades, centuries, millions or even billions of years are being reviewed or projected, it is an empty chant. The Earth is not a stable habitat, nor is the universe around it, and we are not a stable species.

The traveler Freya Stark contrasted a life of insulated safety within the highly entailed techno-complex of urban civilization with the simplicity and expedience of the Bedouin tents where she was welcomed as an honorary man: "Every wild animal lives in a state of danger, which means deciding all the time, while the very essence of tameness is the absence of any need for decision."[12] Her concern, that the benefits of civilization cut us off from a sense of direct causality, is topical as we sit in our houses, glued to screens, following the latest developments in climate-change politics.

It seems harsh to deride as evolutionarily backward Tasmanians who were so confident of their ability to organize critical warmth that they didn't panic when things went wrong, knowing that social mechanisms would guarantee a light, even from their enemies. Looking at the politics of the Eurasian gas pipelines running through Azerbaijan, Ukraine, and Russia to supply western Europe, I wonder if modern Britain has such good guarantees. In the United States, increasing dependency on imported oil, with no currently viable alternative technologies (despite biofuel), is identified by ABC News as "the nation's Achilles heel." Our heat and light could run out. Our enemies might not rush to offer aid.

Long term, I wonder which is more sustainable and resilient, the native Tasmanian way of living in Tasmania, or the modern one? In barely two centuries the island landscape has become scarred by overgrazing, mine-working, and logging, and has hardly been tested at all by the sort of major climate changes that we know the Aboriginal tribes came through with credit.

It is becoming accepted by climate scientists that we are entering a new period of Earth history, the *Anthropocene*, in which our technology has emerged as the most critical factor impacting on global environment. Although we did not plan it, it may be a good thing. Before, we were at the mercy of successive ice ages, as changes in the sun's activity and the Earth's orbit conspired to bring about periodic phase shifts. It was in the last of these that our species evolved, as rapid cooling and drying in East Africa produced new habitats, with opportunities for newly bipedal apes to succeed. This large-scale Ice Age, starting coincidentally at around the time the first stone tool was chipped, has contained lots of individual periods of glacial advance and recession, such as the "last Ice Age" in the Northern Hemisphere, another of which, considered in terms of natural systems (the interplay of Systems 1 and 2), might be expected. A return of the ice would challenge the power of our technology to adapt, perhaps to the breaking point, given current world population levels. But the onset of the Anthropocene, characterized by the high levels of atmospheric carbon-dioxide generated by System 3, means that may never happen again.

Viewed in the perspective of the climate changes of the last 40,000 years, the Tasmanian culture was fairly sustainable. Having arrived during a cold spell, when enough water was locked up in ice to link the land to mainland Australia, the tribes adapted as the weather warmed, paring their technology down through logical choices until it was the one observed historically by Europeans. There is no reason to suppose that, faced with another period of climatic upheaval, the Aboriginal Tasmanians, being smart bipeds and artificial apes, would not have continued to adapt and survive, evolving their technology to suit. By harvesting seals rather than fish, they protected and encouraged fish stocks around the coasts; by not building houses or clearing trees for agriculture, they were ensuring the forest in perpetuity.[13] In this perspective, the last 200 years of mineral extraction, logging, and sheep-farming, with their various and widespread environmental impacts, looks like it might be far less sustainable.

Whether such considerations played any conscious part in Aboriginal Tasmanian reckoning is questionable. It would be a mistake to save the Tasmanians from racist slanders and inappropriate cultural judgments only to hold them up as ideal, noble savages, living timeless, perfectly adapted lives. In fact, in Australia generally, the loss of many of the larger wild animal species may have been the result of an initial orgy of unregulated, unsustainable hunting. The same may have been true in the Americas. The bands and tribes that European colonists encountered in each place may have been the descendants of those who had so degraded their environments that, without some outside impetus—such as the introduction of domestic livestock—options had become limited.

Left in isolation, the Aboriginal Tasmanians would have sustained themselves for as long as they were able, but without suitable animals to domesticate, they would have remained as hunter-gatherers. Without farming, they would not have become sedentary, and it is most unlikely that they would have developed—through the emergence of ceramics—the pyrotechnic knowledge needed to understand the production of metals. And, without an urban life based on farming, there would have been no pressing need for writing beyond the simplest mnemonic symbol systems.

Those activities are part of our current technological development, brought to Tasmania by the Europeans two centuries ago. The short period that has elapsed since has taken us from the wooden-hulled *Beagle*, via the ironclad *Truganini*, to the Saturn V rocket (not built in Tasmania, but part of the entailment of global technology). And that is critical because, in the far longer term, if we do not eventually move off the surface of our planet, or find means to stabilize our local star so that it does not become a red giant, none of it is sustainable. Once the sun's relatively genteel fluctuations cease and it begins its final expansion, blow-torching our atmosphere off into space, the only way out will be up, both metaphorically and literally. Long before the point of ultimate discomfort, we—or our evolutionary descendants—will have needed to trade on our expertise to find ways of surviving elsewhere in the solar system. Perhaps, even, beyond it.

The problem we face may not so much be with the things we cannot control, like the behavior of the sun, but the things that we could control, but do not. I fear that what Sigmund Freud once said of children—"What a distressing contrast there is between the radiant intelligence of the child and the feeble mentality of the average adult"[14]—may also be true of technological culture as it grows to maturity. The cycles of the rise and fall of civilizations, intensively studied by archaeologists over the past decades, show some repeated patterns. One is that civilizations are vulnerable to environmental degradation of their own making, and may intensify their activities in precisely those areas that cause the worst problems as the crisis worsens. Another is that, because new technologies create wealth and power, they are not politically neutral; vested interests become attached to them, and this then stifles further innovation, leading to a kind of stagnation where smart solutions to the environmental problems cannot be freely developed.

Technology, especially the baby-carrying sling, allowed us to push back our biological limits, trading in our physical strength for an increasingly retained infantile early helplessness that allowed our brains to expand, forming themselves (both phylogenetically and ontogenetically) under the increasingly complex artificial conditions of System 3. In terms of brain growth, the high-water mark was passed some 40,000 years ago. The pressure on that organ has been off ever since we started outsourcing intelligence in the form of external symbolic storage. That is now so sophisticated through the new world information networking systems that what will emerge in future may no longer be controlled by our own volition.

Which may be a good thing—it is the unexpected edge-effects of technology that often turn out to have the most potential for the future. By transforming part of a mammoth into the semblance of a human head, the carver of the Brno puppet was honing a sensibility that would eventually allow humans to define the geometry of spheres and visualize the shape of our home planet. As physical objects, such perfected balls, now made of elephant ivory, provided recreation for Charles Darwin in his favored billiard room at Down House. Meanwhile, in Africa, pressure to kill the largest bull elephants to supply raw materials for the popular game inverted the usual natural selection logic. The resultant collapse in herd numbers led, via a skeuomorphic development, to the first industrial plastic, nitrocellulose. The celluloid balls looked enough like the ivory ones to pass muster, but chemical volatility in their production encouraged the development of alternative synthetic compounds, which then began to be put to uses beyond the game table. Smaller red and white plastic balls, joined by rods, visually conceptualized the underlying nature of the atomic manipulations that allowed synthetic products to be designed. Eventually, arranged in a double helix pattern, they would reveal the basic building block of Darwinian natural selection. In parallel, the development of durable low-friction bearings facilitated the perfection of Asimo, the robotic descendant of the Brno puppet master.

Earlier on I cited the philosopher John Gray saying that "It may be in their capacity for consciousness that humans and the machines they are now devising are most alike." We are both product and producers of the third system, the realm of artificiality. It has a capacity to elide time and endure in physical form indefinitely. It could also destroy our planet. But there is no back-to-nature solution. There never has been for the artificial ape.

ACKNOWLEDGMENTS

I OWE MANY DEBTS OF gratitude to those who have helped me frame the ideas in this book, and who have practically helped me complete it, but I will be brief. First, to my wife, Sarah Wright, with whom I edit a journal, who is always my first and last reader, and the person with whom ideas are discussed, modified, and rejected; her passion for Russian travelogue, for Chekhov, Semenov, de Custine, and others, is reflected in these pages. Keeping it in the family, my daughters, to whom this book is dedicated, graciously allowed their images to be used; Josephine contributed many constructively critical comments, as did Rebecca, who also prepared much of the bibliography.

Among academic and professional colleagues, my thanks for discussions on a wide range of technical issues are especially due to Everett Bassett, Peter Hiscock, Glen Jeffrey, Anton Kern, Julia Lee-Thorp, Chris Ruff, Graeme Swindles, and Gerhard Trnka; where I have not followed their guidance it is not their fault. Those involved in cave excavation with me have also made significant contributions in supporting the research agenda that is reflected in these pages, especially John Howard, Ben Neil, Garry Rushworth, Niger Spicer, Neville Steed, John and Wendy Thorp, and Robert White. The graphic diagrams were designed by Tom Frankland, and benefited from discussion with Pat Hadley.

I have also benefited greatly from discussions with, and reading the work of, Katherine Denning, Andy Jones, Carl Knappett, and Natalie Uomini. For sterling support with the photographs, I thank Anton Kern and Andreas Anwora, and Anton Kern too for unique opportunities to participate in Hallstatt fieldwork and to work on the Venus of Willendorf. I am grateful to academic staff at the Brno museum for the opportunity to examine the Brno II puppet. In helping me think more clearly I am deeply grateful too to non-archaeological friends: Rondi Brower (and her colleagues at Blackwood & Brouwer), Phil Brown, Hugh Cornwell (and all the staff at Cheddar caves), James Cox, Keith Moe, Christopher Potter, Daniel Richler, Tina Tau, Gwyneth Vernon, Helen Wheatley, Ben Wheatley, and Ollie Wright. The hardest constituency to please, and yet always supportive, has been my students, past and present. You know who you are, but I should particularly like to mention Holly Carter Chappell, Mhairi Maxwell, and Karin Weinhandl.

Bringing the ragged ideas together, from proposal to page proof, my deep thanks go to Katinka Matson and John Brockman at Brockman, and at Palgrave Macmillan to Luba Ostashevsky, whose critical input (as well as patience), along with that of her editorial colleagues, Erica Warren and Laura Lancaster, has been world class.

NOTES

Introduction

1. Ashmolean Museum 2002; the opening of this chapter, as well as the title of the book, pays homage to Desmond Morris's 1967 classic *The Naked Ape*. Humans are here considered as being one of the great apes in the spirit of kinship rather than strict biological taxonomy.
2. The happenings described are gleaned and inferred from a number of sources: for Tradescant, see Ashmolean Museum 2002; Eija-Riitta Berliner-Mauer's own website about her sexual obsession, marriage, and widowhood when the Berlin Wall was pulled down is at www.berlinermauer.se; the 650,000-year-old traces of scalping are from Bodo, Ethiopia (White 1986); for the cross-dressed terrorist suspect, see Melanie Phillips 2006; a good background to French Tasmanian relations at first contact is given by John Mulvaney 2007 (see also epress.anu.edu.au/aborig_history/axe/mobile_devices/index.html); the limestone chunk is the Venus of Willendorf (Taylor 2006b, 2008); porn star Mary Carey (whose similarity in both name and appearance has caused a certain pop star some grief) is the subject of much web gossip over her breast implant auction—see Katherine Gammon's report for ABC News (Gammon 2007); the soldier was the war poet Keith Douglas, describing his own sniper's experience in "How to Kill" (1943; see various anthologies and www.poemhunter.com/poem/how-to-kill); the scissors belonged to the Avebury barber surgeon (see Denison 1999); the hill is Silbury Hill, and the estimate of the time taken to build it is from Atkinson 1974, p. 128; the site is Gona, in Ethiopia, where the earliest dated chipped stone tools have been found (Semaw et al. 2003); the cyber neglect is at "Internet Obsessed Pair 'Let Baby Die,'" *The Daily Telegraph*, March 6, 2010, p. 23; the scenario at 3.5 million years ago is imagined, but relates to the *Australopithecus afarensis* footprint trails found at Laetoli, Tanzania (Leakey and Hay 1979).

3. Tim Ingold (2001b) provides a useful overview of the way that different Greek and Roman terms for skill or craft once coincided, rather than being separate as they often appear in our modern arts vs. technology division.
4. Kurzweil 2005.
5. Kelly 1994, 2007.
6. Just as System 2 must respect the laws of System 1, even if its emergent powers subvert it (as when life acts against entropy, for example), so System 3 will not easily free itself from System 2.
7. Gray 2002, p. 188.
8. On the edge of the Irish Sea at Corby (see figure 16 on p. 157).

Chapter 1

1. Darwin 1997, p. 267.
2. The most complete initial account is Spindler 1994; key recent analyses include Rollo et al. 2002, Müller et al. 2003, and Ruff et al. 2006; Ötzi has somehow become an introductory leitmotif for me, discussed at the beginning of both *The Prehistory of Sex* (in connection with modern gender biases in interpretation) and *The Buried Soul* (in relation to the reconstruction of past worlds of belief).
3. For a good general account of European—and global—prehistory, see Scarre 2009.
4. The issue of intermediate species between these two, such as, variously, *H. antecessor*, *H. cepranensis*, and/or *H. mauritanicus*, in Europe and Africa, is complex and controversial, with several different schools of thought.
5. From Gona: Semaw et al. 2003.
6. At Dmanisi: Lordkipanidze et al. 2006.
7. *Sahelanthropus tchadensis*: Guy et al. 2006.
8. It should be noted that Kivell and Schmitt 2009 have recently argued that upright-walking hominins cannot have evolved from a knuckle-walking species; the implication is that the common ancestor of chimpanzees and humans is likely to have been arboreal (tree-swinging), with knuckle-walking and bipedalism developing in separate directions thereafter; Kivell and Schmitt also argue that knuckle-walking evolved independently in gorillas and chimps; the trend of this interpretation fits well with more general arguments about speciation and independent parallel evolution made by Venditti et al. 2010.
9. Secord 2008; see James Secord's introduction to this selection of Darwin's key writings.
10. There is a huge and complex literature on this; a good starting point is Weiss et al. 2009 ("Social and Ecological Benefits of Restored Wolf Populations," at www.wildlifemanagementinstitute.org).

11. White 1959, p. 8.
12. Eliot 1867, vol. 4, p. 377.
13. Spencer 1864.
14. Secord's 2008 selection is a good starting point, but many editions of this classic 1859 work exist; Darwin has a lively and engaging style that makes the work a continuing good read.
15. The domestic background and the intellectual ramifications are well described by Adrian Desmond and James Moore in their monumental biography, *Darwin* (Desmond and Moore 1991).
16. Scarre 2009 provides detailed background on all these sites; K. Kris Hirst provides an excellent short introduction to the "History of the Domestication of Cattle" on the archaeology page of the About.com guide; key technical references include Bradley et al. 1998 and Götherström et al. 2005.
17. Mason 1996.
18. See Paul 2009 for a discussion of Galton's "eugenics" (a term he coined).
19. "Meme" was a term coined by Dawkins 1989 and elaborated by Blackmore 1999; it is discussed in greater detail later.
20. Ruff 2002.
21. See Ruff 1994 for the importance of Bergmann's Rule in human evolution; see Allen 1887 for Allen's rule and also Hurd et al. 2007.
22. See contributions in Brantingham et al. 2004.
23. Brian Fagan's 2004 edited volume *The Seventy Great Inventions of the Ancient World* provides a good overall guide to what was invented when, at least in terms of surviving hard evidence (there are few significant inferences about organics and perishables, and slings are mentioned only in relation to later prehistoric warfare rather than earlier prehistoric hunting and infant transport).
24. Kern et al. 2008, p. 61ff; 2009.
25. Mollison 1994.
26. Taylor 1996, 2006c.
27. Churchill made this comment in a wartime radio broadcast on March 21, 1943 (*Complete Speeches*, 1974, vol. 7).
28. Wrangham 2007, 2009.
29. Ruff 1994.
30. Ruff et al. 2006.
31. I phrase this rhetorically; animal intelligence, understood as sensory knowledge of the environment, may frequently be superior to ours. But animals are stupid in contrast to humans when we turn to what are called (by us) the higher cognitive functions.
32. A range of statistics can be found at the National Eye Institute's website, www.nei.nih.gov/CanWeSee.
33. Darwin's journal for December 25, 1832 (Darwin 1997, p. 203).
34. Darwin's journal for December 17, 1832 (Darwin 1997, p. 199).

35. Darwin writes without demur that "the different tribes then at war were cannibals" (Darwin 1997, p. 204), but one of his modern editors, Secord (who is also keen to assure us that the naked nursing mother in the sleet had a diet rich in meat and blubber that "afforded considerable protection against the cold" [Secord 2008, p. 438, note 23]), cites "unequivocal" anthropological literature (source unidentified) that would deny cannibalism (op. cit., note 24); and ditto Desmond and Moore (2009, p. 96), who write of "a gullible Darwin" who believed the Fuegian "joke" about eating relatives during famines (of course, their picture of Darwin is sure-footed enough not to render him typically gullible at other times). I have discussed "visceral insulation" and the ill-informed, revisionist, inverted imperialism of anti-cannibal social anthropology *in extenso* elsewhere (Taylor 2002); but, in short, there is no *a priori* reason to doubt the plausibility of what Darwin records he was told.
36. Brown 1987; for the broader context, see also Lieberman et al. 2002; Lewin and Foley 2004.

Chapter 2

1. This appears in his journal entry for Tierra del Fuego, December 17, 1832 (Darwin 1997, p. 196); it is not in any sense a throwaway line as in October 1836, finally nearing home, he repeats the passage, almost verbatim (except "tame" replaces "domesticated"), and prefaces it with an early indication of its relevance to broader human evolution: "perhaps nothing is more certain to create astonishment than the first sight in his native haunt of a barbarian—of man in his lowest and most savage state. One's mind hurries back over past centuries, and then asks, could our progenitors have been men like these?" (Darwin 1997, p. 477). Of course, Darwin's views on race, humanity, and evolution changed over time (see Hodge and Radick 2009 for a thorough introduction to the development of Darwin's thinking); but Desmond and Moore's 2009 *Darwin's Sacred Cause*, in seeking (correctly) to demonstrate Darwin's powerful antislavery credentials alongside his (demonstrably correct) single-species concept of humans, underplays the extent to which it was his moral sensitivity that informed his position on human-human exploitation, not an acceptance of any biological equity between civilized and savage (in his terms); he hated cruelty to animals as well as people, but he also believed that evolution was endlessly active among us, pitting weaker and fitter individuals, but also varieties, of a single species against one another in the battle for survival.
2. Quotations here are from Darwin's *Voyage of the Beagle* journals (Darwin 1997, p. 208); see previous footnote. Darwin's first impressions provide a powerful insight into the way primitive peoples were viewed by nineteenth-century Europeans precisely because Darwin was neither xenophobic nor

at any time unclear that he was dealing with one and the same species; of course, his massive, enduringly destabilizing contribution was to undermine "species," and therefore categorical (monothetic) definitions of humanity.

3. De Strzelecki 1845, p. 334.
4. Darwin 1997, p. 411.
5. The quotation is from chapter 6, "On the Affinities and Genealogy of Man," pagination varying according to edition, and curiously omitted from Secord's 2008 selection of Darwin's *Evolutionary Writings* (Secord 2008, p. 272: square ellipsis and gloss between chapters 5 and 7; "Darwin examines the place of humans...and concludes with a speculative genealogy"). The easiest-to-find source is the admirable Project Gutenberg (www. Gutenberg.org); the original reference is Charles Darwin, *The Descent of Man* (1871, 1st ed.), p. 168f.
6. I have not consulted Klemm in the original yet: a summary of his views from his ten-volume *Allgemeine Cultur-Geschichte der Menschheit* (Universal History of Human Cultures), 1843–1852, and the *Allgemeine Kulturwissenschaft* (General Science of Culture), 1854–1855, is presented by Trigger 2006, p. 168.
7. Duyker 1992; see also George Weber's useful www.andaman.org.
8. De Strzelecki 1845, p. 350.
9. The quotations are from Margaret Sanger's radical sexual health guide, *What Every Girl Should Know* (Sanger 1920), which betrayed both radical early feminism and a sharply racialized eugenic sensibility, gleaned from works such as Alexander Sutherland's *Victoria and Its Metropolis*, from whence the second uncomfortable sound bite.
10. Quotation from Bonwick 1884, p. 217.
11. *Encyclopaedia Britannica* 1919, vol. 21, p. 829. "Tasmanians," a people described as "now extinct [with] longish, oval or pentagonal, flattish and small sized (cranial contents) heads; short broad noses and large teeth." The entry is anonymous but could potentially have been written by Richard Berry (see next note).
12. The view of the anatomist Richard Berry, as recorded in 1907; Berry was later president of the British Medical Association and published (with Robertson and Büchner) a work on Tasmanian Aboriginal craniometry (Berry et al. 1914).
13. De Strzelecki 1845, p. 334.
14. It must be said that archaeological work now clearly shows a long tradition of adaptation to changing environmental conditions (e.g., Cosgrove 1999; Hiscock 2006, 2008), but my rhetorical phrasing here is meant to indicate a contrast between communities capable of periodic change and new arrivals for whom change was centrally culturally valorized.
15. Of these, perhaps Petr Petrovich Semenov is the least known in Anglophone circles; a good start would be Colin Thomas's 1998 English-language edition of *Travels in the Tian'-Shan' 1856–1857*.

16. Hiscock 2008, chapter 7, now provides the standard account; my thinking about the fish issue owes much to Everett Bassett (see Bassett 2004).
17. McGrew 1987; McGrew was following a line of argument first elaborated by Sollas in 1911 (see discussion in Cosgrove 1999, p. 359).
18. Diamond 1997, especially chapter 13. See also Diamond 1993, 1998.
19. The following series of quotations is from Diamond's preliminary 1993 article for *Discover* magazine, "Ten Thousand Years of Solitude" (Diamond 1993 and available online at www.discovermagazine.com/1993/mar/tenthousandyears189).
20. Jones 1977a, p. 197; see also Jones 1972, 1995.
21. Jones 1978, as cited by Diamond 1993.
22. Jones 1977b, p. 203.
23. In *Lord of the Flies*, Golding imagines a party of schoolboys reverting from civilized values to primitive savagery when isolated; the imaginative *Lord of the Flies*–style hypothesis concerning the Tasmanian fish taboo comes from Tim O'Neill/Thiudareiks Gunthigg on the Total War Center site at www.twcenter.net/forums.
24. Sim 1999.
25. Bassett 2004.
26. Colley and Jones 1987; and see the discussion in Hiscock 2008, p. 133f.
27. Gott 2002.
28. Cf. Hiscock 2008, p. 136.
29. Chekhov 1895, chapter 11, translation from Chekhov 1989, p. 103.
30. Hughes 1987, p. 424, provides one of the best brief contextualized accounts.

Chapter 3

1. Houellebecq 2005; translation here by Sarah Wright.
2. Lewin and Foley 2004.
3. Paley 1802 (see Paley 1996).
4. Paley 1837, p. 55. I have omitted some emphases and reference to plates.
5. Paley 1837, p. 58.
6. Secord 2008, dust jacket.
7. See www.janegoodall.org/chimp-central-pets; Wallace 1856.
8. Mullen 2005.
9. Lynn Kilgore's work is summarized in Lovell et al. 2000.
10. See Hart and Sussman 2008 for the most up-to-date summary.
11. Desmond and Moore 1991.
12. The phrase was used several times in *The Descent of Man* (see Darwin 2004, p. 502).
13. Darwin 2004, p. 513.

14. *Real Vampires*, Discovery Channel, 2007 (dir. Daniel Richler).
15. Dorion 2004, p. 7.
16. Stoker 1897 and many later editions, chapter 8.
17. Large numbers of different bite-force claims are made on a plethora of websites devoted to fighting dogs, their fans, and detractors; there appears a corresponding absence of statistically valid and controlled scientific measurement.
18. *Dracula*, 1931 (Universal Pictures; dir. Tod Browning).
19. Berger and Clarke 1995.
20. Carroll 2005, p. 272ff.
21. There is a massive literature on potential anthropogenic aspects of the extinction of megafauna on various continents; Alroy 2001 is a good starting point for North America.
22. Dembski and Ruse 2004.
23. See Calvin 1991; Maslin and Christensen 2007.
24. Johanson and Maitland 1981.
25. However, this does not vitiate their immense importance in advancing our understanding of human origins; see also Johanson 2009.
26. Daniel 1967 provides a good account with a transcription of the letter.
27. Daniel 1967.
28. Daniel 1962, p. 42.
29. By Grahame Clark; see Foley and Lahr 1997 for a detailed discussion.
30. Aubry et al. 2009.
31. Deino et al. 2006; Prat 2007; Aubry et al. 2009.
32. Aubry et al. 2009.
33. This is not exact, of course, but of the correct order of magnitude.
34. Martin Pickford and others have recently published a number of technical papers on this complex field.
35. But it may be that a variety of niches were still able to be exploited; see Sponheimer and Lee-Thorp 2003.
36. Wilford 2007.
37. Asfaw et al. 1999.
38. Lewin and Foley 2004.
39. Bernard Wood and Mark Collard are among those who see strong australopithecine features in what is conventionally termed *Homo habilis*.

Chapter 4

1. Giles in Short 2003, p. 167f.
2. Short 2003.
3. Wrangham 2009.
4. Wrangham 2007.

5. Wrangham 2009, jacket.
6. Karlinsky 1973.
7. A widely quoted judgment.
8. A widely anthologized statement.
9. Lévi-Strauss 1970.
10. Douglas 1975, p. 244.
11. Ibid.
12. Cochrane 1829 (see Cochrane 1983).
13. Cochrane 1983, p. 128.
14. Cochrane 1983, p. 129.
15. Cochrane 1983, p. 129f.
16. David 1965, p. 28.
17. Leroux 2007.
18. www.carnegieweightmanagement.com.
19. *Michael Phelps' diet (12000 calories a day)*, www.youtube.com.
20. Aiello 1997.
21. Wrangham 2009.
22. See Bateson 2004 for a good introduction and review.
23. However, this is disputed, with some claiming that Inuit diets were higher in protein than fat; problems with dietary change in recent decades make this hard to investigate.
24. Chekhov 1989.
25. Taylor 2002, chapter 3, for a general introduction.
26. Tanner 1994 is available but heavily edited; early editions are rare books.
27. Tanner 1830.
28. See Taylor 2002.
29. On the attraction side, one should note that many animals, including primates, have highly selective SMRS (Specific Mate Recognition Systems).
30. As well as purely terminological arguments, the discoveries of genomics will continue to change the phylogenetic picture.
31. Stringer and McKie 1997.
32. Darwin 1997, p. 413.
33. Darwin 1997, p. 196; this view is nuanced to some degree, but not, to my mind, altered, in the discussion in Desmond and Moore 2009.
34. Darwin 1997, p. 204.
35. Ibid.
36. Ibid.
37. Secord 2008, p. 438, note 24.
38. Desmond and Moore 2009, p. 96.
39. Taylor 2002, p. 273.
40. www.arch.cam.ac.uk/projects/tellbrak.
41. See Roaf 1990, p. 65, for a picture; the quality is discussed further in chapter 7.

42. Barnes, Duda, Pybus, and Thomas, forthcoming.
43. Taylor 2002.
44. This is another aspect, in fact, of visceral insulation.
45. "Eaten Missionary's Family Get Apology," http://news.bbc.co.uk/2/hi/asia-pacific/3263163.stm.
46. The initial apologies, one of which involved an attempt to return the boots, were not accepted; it seems that part of the motivation in persisting was a feeling that trade, tourism, and modern communications technology were bypassing the island—a curse consequent on having eaten the missionary.

Chapter 5

1. Hager 1997, p. 4.
2. Köhler and Moyà-Solà 1997.
3. Just being able to walk a few steps in the upright position is not enough. Gorillas, for example, can be bipedal for a short while, but this is not their habitual mode of locomotion.
4. Calvin 1991; see also Knüsel 1992.
5. Daniel 1967.
6. Bogin 1997; Falk 2009.
7. There is a long-standing debate about the status of sexual selection, with some arguing that it sensibly reduces to natural selection; one of the problems with it is how to measure fitness, which is why proxy measures, such as immunity, are used.
8. In nature at least; cf. an increasingly techy world, where mental incapacity *might* be disadvantageous, at least at a group level.
9. Johanson 2009.
10. Ruff 1995, 2010.
11. See discussion in earlier chapters and the extensive detail in Desmond and Moore 1991. On these issues see also works by Robin Dunbar, and Christopher Wills's 1993 monograph on sexual selection driving human intelligence.
12. Hager 1997.
13. Hager 1997, p. 6.
14. See Zihlman 1997 for a short history.
15. Statistics from Cumbrian mountain rescue services; the advent of GPS and mobile phones seems to have increased the number of emergency calories, as people go walking without map or compass (I discuss this in greater depth in chapter 8).
16. McKiernan and Lieberman 2005.
17. Wrangham 2009, p. 131.
18. The excursion was part of the 1990 SAA meeting in Las Vegas.

19. Ehrenberg 1989, p. 47f.
20. Bolen 1992, p. 54.
21. Falk 2009, p. 51.
22. Wall-Scheffler et al. 2007.
23. Ibid.
24. Ibid.
25. As reported in Pseudo-Plutarch, *Miscellanies*, 179.2; see Graham 1999.
26. As reported by Censorinus, *On the Day of Birth*, 4.7.
27. This maxim was made famous by Ernst Haeckel.
28. On this theme, see Gould 1977.
29. Gould 1979.
30. Darwin 1872; for a modern image Darwin would have almost certainly approved of, Google "happy-puppy-totally-looks-like-happy-baby."
31. Portmann 1941.
32. Purves and Lichtman 1985.
33. Tattersall 2007, p. 137.
34. Bard 1995.
35. Hill et al. 2001.
36. Brain size as a raw measure can be misleading, as brains vary according to body mass. Normalizing for this gives the "encephalization quotient" (EQ), which may be taken as a better proxy for intelligence, but not absolutely so: a mouse with a very large brain relative to its body would still not be as smart as a human; in addition, the relative sizes of different parts of the brain are also important, with the development in humans of the prefrontal cortex being especially marked.
37. Carbonell et al. 2008.
38. Berger and Clarke 1995.
39. Klein and Edgar 2002, p. 87f.
40. The more it becomes apparent that, for long periods in our evolution, several species of hominin existed at broadly the same time in the same location, the harder it becomes to ascribe a particular technology to one group; given the presence of mimicry in primates, and that groups take things from one another, once chipped stone tools start getting made—by whomever—they may in practice start being used by more than one species.

Chapter 6

1. Carlyle was describing what he called "The 'Mechanical' Age" (Carlyle 1858).
2. Bahn and Vertut 1997.
3. Chauvet: Clottes 2003; El Pindal: Balbin Behrmann et al. 1999.

4. See Renfrew 1994, 2001, 2004; Renfrew and Zubrow 1994; Renfrew and Scarre 1998; see also Mithen 1996, 1998.
5. Zihlão et al. 2010.
6. The term was coined by John Pfeiffer (Pfeiffer 1982).
7. Schlanger 1994.
8. Leroi-Gourhan 1964.
9. Soffer et al. 2000; Adovasio et al. 2007.
10. In interview with Angier 1999.
11. Taylor 2008.
12. Delporte 1993; Taylor 2008.
13. Soffer et al. 2001; Viegas 2000.
14. Illingworth as cited in Viegas 2000.
15. Dawkins 1983, 1989.
16. Dawkins 1989, chapter 11.
17. The development of the idea has gone ahead, especially with Boyd and Richerson 2000; Blackmore 1999; Conte 2000; Laland and Odling-Smee 2000; Shennan 2002.
18. My concerns are shared by many: Bloch 2000, Sperber 1994, 2000.
19. Recent works, ultimately following from early forays such as Childe 1936; Bourdieu 1970; and Jaynes 1976 and the foundational work of Godelier 1985; Appadurai 1986; Kopytoff 1986; Lemonnier 1986, 1992; Miller 1985; and Donald 1991, there has been a vibrant literature, including Boivin 2004; Brown 2001; Chapman and Gaydarska 2006; Clark 2008; Damasio 2000; DeMarrais et al. 2004a, 2004b; Donald 1991; Gell 1992, 1998; Gibson and Ingold 1993; Gosden 1994, 2004; Ingold 2001a, 2001b; Johnson 1993; A. Jones 2007; Knappett 2002, 2004, 2005, 2008, Knappett and Malafouris 2008; Latour 1999; van der Leeuw 2008; Malafouris 2004; Robb 2004; Schiffer 2001; Schlanger 1996.
20. Goody 1979.
21. He continues to be hugely productive.
22. See www.adbonline.anu.edu.au/biogs. NB: The *Australian Dictionary of National Biography* cautions that the first names of du Fresne are often wrongly given, as Nicolas-Thomas or Marie-Joseph.
23. Weber: http://www.andaman.org/BOOK/chapter52.
24. Rousseau's remark is so widely reported that its precise context is no longer clear; Brian Fagan has Rousseau saying this exclusively of the cannibal Maoris (Fagan 1997, p. 72). But the disillusioned sentiment is clear enough.
25. Marquis de Custine, Petersburg, July 10, 1839 (2001, p. 44).
26. Deane 1967, p. 84ff.
27. Cosgrove 1999.
28. Babel, *The Story of My Dovecote*; translation by Sarah Wright.

29. Mauss 1990.
30. Valoch 2009.
31. That the Czech Republic should have produced the Brno puppet, the Puppet Theatre of Prague, and the playwright Karol Čapek, whose *R.U.R.* play first introduced the concept of the robot to a wide audience is, I guess, chance.

Chapter 7

1. Pratchett 1988.
2. Morton 1994, p. 125. (Wivenhoe is not fictional, of course; Morton chose places like this picturesque Essex village to gently lampoon English small-town eccentricities.)
3. Gell 1998, p. 101.
4. Sherratt and Taylor 1997.
5. Ibid; the Ice Man may have died as early as 3,350 B.C. and as late as 3,200 B.C. Due to a wiggle in the radiocarbon calibration curve at this point, it is hard to be more exact, but later in the bracket may be marginally more likely.
6. Information from the anonymously authored *Die gute alte Küche Wien* (Ueberreuter 2006).
7. The performance referred to is the Royal Shakespeare Company's 2008 production, as filmed by the BBC and screened on December 26, 2009.
8. RSC press release.
9. Gormley's *Event Horizon* has since appeared in New York.
10. Melville, *Moby Dick, or the White Whale.*
11. This is a very knotty philosophical problem.
12. Clarke 1968.
13. On the Gundestrup Cauldron; the topic is not especially relevant for the discussion here.
14. Taylor 1996, 2006c.
15. Schaefer 2002.
16. Chekhov 1989.
17. Johnson 2002 and personal communication.
18. Detail provided in the *Australian Dictionary of National Biography*, adbonline.anu.edu.au/adbonline.htm, under "Girardin."
19. adbonline.anu.edu.au/adbonline.htm.
20. *Sunday Tasmanian*, January 16, 2005.
21. "Bye-bye, Sweet Bay," www.eniar.org/news/Recherche.html.
22. De Strzelecki 1845.
23. Ibid.
24. Actually except Thailand, which caters to a much more international clientele ("Iran's Diagnosed 'Transsexuals'" report for BBC News by Vanessa Barford, February 25, 2008).

25. These photographs are widely available on web-based resources.
26. See chapter 2.
27. Andrys Onsman, "Truganini's Funeral," http://www.islandmag.com/96/article.html.
28. "Captain Sarah's Odyssey: The Hobart Citizen of the Year" (1998), www.abc.net.au/austory/archives/AustoryArchivesIdx_Monday23September2002.htm.

Chapter 8

1. Pfaffenberger 1992, p. 282.
2. Maeve Kennedy, *The Guardian*, July 9, 2003. It should be made clear that Professor Stone did not write that she *wanted* the looters dead, and her concerns were articulated at an academic conference, held in the British Museum, on the problem of Iraq's heritage following the toppling of Saddam Hussein, where her words were reported by a diligent journalist.
3. J. Cheng, "Bevel-Rimmed Bowls," World History Sources website, http://chnm.gmu.edu/worldhistorysources, accessed March 2010 (Center for History and New Media, George Mason University).
4. Or *ninda*, bread; Roaf 1990, p. 70.
5. See Roaf 1990, p. 65, for an image of a BRB and p. 70 for a guide to the bowl and head glyphs and their phonetic values.
6. Roaf 1994.
7. "Photos Document Destruction of Afghan Buddhas," CNN, March 12, 2001.
8. Belting 1994, p. 1.
9. Belting 1994, p. 9.
10. Book III, xvi-xxiv.
11. Crawford 2007, p. 27.
12. Byatt 1998, p. 51.
13. On Wikipedia commons, if you search for historical pictures of the Tasmanian Museum.
14. See www.seti.org. Frank Drake is currently board chairman of the SETI Institute; an interactive version of the equation can be found at www.activemind.com/Mysterious/Topics/SETI/drake_equation.html.
15. See Taylor section of SETI podcast, *Driving Evolution*, 2007, with Seth Shostak.
16. Gomes et al. 2005.
17. Schidlowski 2001.
18. Hoyle and Wickramasinghe 1981.
19. Estimates made by the University of Arizona cosmologist Paul Davies; a series of detailed figures are given by him in his influential book *The*

Cosmic Blueprint. Davies firmly believes that emerging life in the universe is structured by some underlying "creative," nonrandom process.

20. Dartnell 2007.
21. Flusser 1999.
22. www.afghanculturemuseum.org.

Conclusion

1. Kelly 2007, www.edge.org/3rd_culture/kelly07/kelly07_index.html.
2. Gellner 1989, p. 513.
3. Gellner 1989, p. 515f.
4. Laland and Odling-Smee 2000; Laland et al. 2010.
5. Chekhov 1989, p. 244.
6. *The Brisbane Courier*, February 24, 1891, p. 5, newspapers.nla.gov.au/ndp/del/page/100591; a Burns Philp & Co. Ltd. postcard image of the ship can be found at www.flotilla-australia.com/burnsphilp3.htm.
7. *The Mercury*, June 19, 1878, p. 2, far left column, http://newspapers.nla.gov.au/ndp/del/page/793247; "Tin-Mining in Tasmania," *Otago Witness*, May 18, 1878, p. 4.
8. Falk 2009, p. 51, and compare pp. 53–55 (in noting this, my aim is not to criticize a deeply fascinating account of language emergence; my uncertainty is at least equal to Falk's).
9. Ruff 2010; Ruff's measurements of Gona certainly leave some doubt about whether the pelvis really does belong to genus *Homo* and not some previously unknown, late surviving australopith.
10. Pasternak 2007.
11. See Garreau at www.edge.org.
12. Stark 1997.
13. For recent work on climate change and the adaptation of human groups to it, see Meltzer and Holliday 2010.
14. Widely anthologized; I have not consulted the original German source for this.

BIBLIOGRAPHY

Adovasio, J. M., Soffer, O., and Page, J. 2007. *The Invisible Sex: Uncovering the True Roles of Women in Prehistory*. New York: HarperCollins.

Aiello, L. C. 1997. Brains and guts in human evolution: The expensive tissue hypothesis. *Brazilian Journal of Genetics* 20(1): 141–148.

Allen, J. A. 1877. The influence of physical conditions in the genesis of species. *Radical Review* 1: 108–140.

Alroy, J. 2001. A multispecies overkill simulation of the end-Pleistocene megafaunal mass extinction. *Science* 292: 1893–1896.

Angier, N. 1999. Furs for evening, but cloth was the Stone Age standby. *New York Times*, December 14.

Appadurai, A. (ed.). 1986. *The Social Life of Things*. Cambridge: Cambridge University Press.

Asfaw, B., White, T., Lovejoy, O., Latimer, B., Simpson, S., and Suwa, G. 1999. Australopithecus garhi: A new species of early hominid from Ethiopia. *Science* 284(5414): 629–635.

Ashmolean Museum. 2002. *The Tradescant Collection* (web resource by D. Berry and J. Moffatt at www.ashmol.ox.ac.uk/ash/amulets/tradescant/). Oxford: University of Oxford.

Atkinson, R. J. C. 1974. Neolithic science and technology. *Philosophical Transactions of the Royal Society of London* (Series A, Mathematical and Physical Sciences) 276(1257): 127–131.

Aubry, M. -P., Berggren, W. A., Couvering, J. van, McGowran, B., Hilgens, F., Steininger, F., and Lourens, L. 2009. The Neogene and Quaternary: Chronostratigraphic compromise or non-overlapping magisteria? *Stratigraphy* 6(1): 1–16.

Bahn, P., and Vertut, J. 1997. *Journey Through the Ice Age*. Berkeley: University of California Press.

Balbin Behrmann, R. de, Alcolea Gonzáez, J. J., and Gonzáez Pereda, M. A. 1999. Une visión nouvelle de la grotte de El Pindal, Pimiango, Ribadedeva, Asturias. *L'Anthropologie* 103: 51–92.

Bard, K. A. 1995. Parenting in primates. In M. H. Bornstein (ed.), *Handbook of Parenting*, vol. 2, *Biology and Ecology of Parenting*: 27–58. Mahwah, NJ: Erlbaum.

Barkow, J. H., Cosmides, L., and Tooby, J. 1992. *The Adapted Mind: Evolutionary Psychology and the Generation of Culture*. Oxford: Oxford University Press.

Bassett, E. 2004. Reconsidering evidence of Tasmanian fishing. *Environmental Archaeology* 9(2): 135–142.

Bateson, P. 2004. The active role of behaviour in evolution. *Biology and Philosophy* 19: 283–298.

Beaune, S. A. 2008. *L'homme et l'outil: L'invention technique durant la préhistoire*. Paris: CNRS Éditions.

Belting, H. 1994. *Likeness and Presence: A History of the Image Before the Era of Art*. Chicago: University of Chicago Press.

Bentley, A. (ed.). 2008. *The Edge of Reason? Science and Religion in Modern Society*. London and New York: Continuum.

Berger, L. R., and Clarke, R. J. 1995. Eagle involvement in accumulation of the Taung child fauna. *Journal of Human Evolution* 29: 275–299.

Berry, R. A., Robertson, A. W. D., and Büchner, L. W. G. 1914. The craniometry of the Tasmanian Aboriginal. *Journal of the Royal Anthropological Institute of Great Britain and Ireland* 44: 122–126.

Blackmore, S. 1999. *The Meme Machine*. Oxford and New York: Oxford University Press.

Bloch, M. 2000. A well-disposed social anthropologist's problems with memes. In R. Aunger (ed.), *Darwinizing Culture: The Status of Memetics as a Science*: 189–204. Oxford: Oxford University Press.

Boden, M. A. 2006. *Mind as Machine: A History of Cognitive Science*, vol. 1 and 2. Oxford: Oxford University Press.

Bogin, B. 1997. Evolutionary hypotheses for human childhood. *Yearbook of Physical Anthropology* 40: 63–89.

Boivin, N. 2004. Mind over matter? Collapsing the mind-matter dichotomy in material culture studies. In E. DeMarrais, C. Gosden, and C. Renfrew (eds.), *Rethinking Materiality: The Engagement of Mind with the Material World*: 63–71. Cambridge: McDonald Institute for Archaeological Research.

Bolen, K. M. 1992. Prehistoric construction of mothering. In C. Claassen (ed.), *Exploring Gender Through Archaeology: Selected Papers from the 1991 Boone Conference*: 49–62. Madison, WI: Prehistory Press.

Bonwick, J. 1884. *The Lost Tasmanian Race*. London: Sampson Low, Marston, Searle, and Rivington.

Bourdieu, P. 1970. The Berber house or the world reversed. *Social Science Information* 9(2): 151–170.

Boyd, R., and Richerson, P. J. 2000. Memes: Universal acid or a better mousetrap? In R. Aunger (ed.), *Darwinizing Culture: The Status of Memetics as a Science*: 143–162. Oxford: Oxford University Press.

Boyd, R., and Silk, J. B. 2003. *How Humans Evolved* (3rd ed.). New York and London: W. W. Norton.

Boyer, P. 1994. *The Naturalness of Religious Ideas: A Cognitive Theory of Religion*. Berkeley: University of California Press.

Bradley, D. G., Loftus, R. T., Cunningham, P., and MacHugh, D. E. 1998. Genetics and domestic cattle origins. *Evolutionary Anthropology* 6(3): 79–86.

Bramble, D. M., and Lieberman, D. E. 2004. Endurance running and the evolution of Homo. *Nature* 432: 345–352.

Brantingham, P. J., Kuhn, S. L., and Kerry, K. W. (eds.). 2004. *The Early Upper Paleolithic Beyond Western Europe*. Berkeley: University of California Press.

Brown, B. 2001. Thing theory. *Critical Inquiry* 28: 1–22.

Brown, P. 1987. Pleistocene homogeneity and Holocene size reduction: The Australian human skeletal evidence. *Archaeology in Oceania* 22: 41–67.

Byatt, A. S. (ed.). 1998. *The Oxford Book of English Short Stories*. New York: Oxford University Press.

Calvin, W. H. 1991. *The Ascent of Mind: Ice Age Climates and the Evolution of Intelligence*. New York: Bantam.

Carbonell, E., Bermúdez de Castro, J. M., Parés, J. M., Alfredo Pérez-González, A., et al. 2008. The first hominin of Europe. *Nature* 452: 465–469.

Carlyle, T. 1858. *The Collected Works of Thomas Carlyle*. 16 vols. London: Chapman and Hall.

Carroll, S. B. 2005. *Endless Forms Most Beautiful: The New Science of Evo-Devo and the Making of the Animal Kingdom*. New York: Norton.

Chapman, J., and Gaydarska, B. 2006. *Parts and Wholes: Fragmentation in Prehistoric Context*. Oxford: Oxbow Books.

Chekhov, A. 1989 [1895]. *The Island of Sakhalin* (translated by L. and M. Terpak; introduction by I. Ratushinskaya). London: Folio.

Childe, V. G. 1936. *Man Makes Himself*. London: Watts & Co.

Clark, A. 2008. Where brain, body and world collide. In C. Knappett and L. Malafouris (eds.), *Material Agency: Towards a Non-Anthropocentric Approach*: 1–18. New York: Springer.

Clarke, D. 1968. *Analytical Archaeology*. London: Methuen.

Clottes, J. 2003. *Chauvet Cave: The Art of Earliest Times* (translated by P. G. Bahn). Salt Lake City, UT: University of Utah Press.

Cochrane, J. D. 1983. *A Pedestrian Journey Through Russia and Siberian Tartary to the Frontiers of China, the Frozen Sea and Kamchatka* (edited by M. Horder). London: Folio Society.

Coghlan, A. 2008. Thank culture for your modern mind. *New Scientist* (May 17): 8–9.

Colley, S., and Jones, R. 1987. New fish bone data from Rocky Cape, northwest Tasmania. *Archaeology in Oceania* 22: 67–71.

Conard, N. J. 2008. A critical view of the evidence for a southern African origin of behavioural modernity. *South African Archaeological Society Goodwin Series* 10: 175–179.

Conkey, M. W. 1997. Mobilizing ideologies: Paleolithic "art," gender trouble, and thinking about alternatives. In L. D. Hager (ed.), *Women in Human Evolution*: 172–207. London: Routledge.

Connerton, P. 1989. *How Societies Remember*. Cambridge: Cambridge University Press.

Conte, R. 2000. Memes through (social) minds. In R. Aunger (ed.), *Darwinizing Culture: The Status of Memetics as a Science*: 83–120. Oxford: Oxford University Press.

Cosgrove, R. 1999. Forty-two degrees south: The archaeology of the late Pleistocene Tasmania. *Journal of World Prehistory* 13(4): 357–402.

Cosmides, L., and Tooby, J. 1994. Origins of domain specificity: The evolution of functional organization. In L. A. Hirschfeld and S. A. Gelman (eds.), *Mapping the Mind*: 85–116. Cambridge: Cambridge University Press.

Crawford, C. L. 2007. Collecting, defacing, reinscribing (and otherwise performing) memory in the ancient world. In N. Yoffee (ed.), *Negotiating the Past in the Past: Identity, Memory, and Landscape in Archaeological Research*: 27–42. Tucson: University of Arizona Press.

Crawford, M. A. 1982. The role of dietary fatty acids in biology: Their place in the evolution of the human brain. *Nutrition Reviews* 50: 3–11.

Cumberpatch, C. G., and Blinkhorn, P. W. (eds.). 1997. *Not So Much a Pot, More a Way of Life* (Oxbow Monograph 83). Oxford: Oxbow.

Custine, Marquis de. 2001. *Journey for Our Time: The Journals of the Marquis de Custine* (edited and translated by P. P. Kohler). London: Phoenix.

Damasio, A. R. 2000. *The Feeling of What Happens: Body and Emotion in the Making of Consciousness*. London: Vintage.

Daniel, G. 1962. *The Idea of Prehistory*. Harmondsworth: Penguin.

Daniel, G. 1967. *The Origins and Growth of Archaeology*. Harmondsworth: Penguin.

Dartnell, L. 2007. *Life in the Universe: A Beginner's Guide*. London: Oneworld.

Darwin, C. 1872. *The Expression of the Emotions in Man and Animals*. London: John Murray.

Darwin, C. 1997. *The Voyage of the Beagle* (reprint of 2nd ed. [1845] of Darwin's *Journal of Researches into the Natural History and Geology of the Countries Visited During the Voyage of H.M.S. Beagle Around the World, Under the Command of Captain Fitz Roy, R.N.*). Hertfordshire: Wordsworth Editions.

Darwin, C. 2004. *Descent of Man and Selection in Relation to Sex* (introduction by H. Cravens). New York: Barnes & Noble.

Darwin, C. 2008. *Evolutionary Writings* (edited by J. A. Secord). Oxford and New York: Oxford University Press.

David. E. 1965. *French Provincial Cooking*. London: Michael Joseph.

Davies, P. 2004a. *The Cosmic Blueprint.* Philadelphia and London: Templeton Foundation Press.

Davies, P. 2004b. Emergent complexity, teleology, and the arrow of time. In W. A. Dembski and M. Ruse (eds.), *Debating Design: From Darwin to DNA*: 191–209. Cambridge: Cambridge University Press.

Dawkins, R. 1983. *The Extended Phenotype: The Gene as a Unit of Selection.* Oxford: Oxford University Press.

Dawkins, R. 1989. *The Selfish Gene* (2nd ed.). Oxford: Oxford University Press.

Dawkins, R. 2005. Untitled essay. In J. Brockman (ed.), *What We Believe but Cannot Prove*: 9. London: Simon & Schuster.

Deane, P. 1967. *The First Industrial Revolution.* Cambridge: Cambridge University Press.

Deino, A. L., Kingston, J. D., Glen, J. M., Edgar, R. K., and Hill, A. 2006. Precessional forcing of lacustrine sedimentation in the late Cenozoic Chemeron Basin, Central Kenya Rift, and calibration of the Gauss/Matuyama boundary. *Earth and Planetary Science Letters* 247: 41–60.

Delporte, H. 1993. *L'image de la femme dans l'art préhistorique.* Paris: Picard.

DeMarrais, E., Gosden, C., and Renfrew, C. (eds.). 2004a. *Rethinking Materiality: The Engagement of Mind with the Material World.* Cambridge: McDonald Institute for Archaeological Research.

DeMarrais, E., Gosden, C., and Renfrew, C. 2004b. Introduction. In E. DeMarrais, C. Gosden, and C. Renfrew (eds.), *Rethinking Materiality. The Engagement of Mind with the Material World*: 1–7. Cambridge: McDonald Institute for Archaeological Research.

Dembski, W. A., and Ruse, M. (eds.). 2004. *Debating Design: From Darwin to DNA.* Cambridge and New York: Cambridge University Press.

Denison, S. 1999. Editorial comment: Lost skeleton of "barber-surgeon" found in museum. *British Archaeology* 48 (October).

Desmond, A., and Moore, J. 1991. *Darwin.* London: Penguin.

Desmond, A., and Moore, J. 2009. *Darwin's Sacred Cause: Race, Slavery and the Quest for Human Origins.* London: Penguin.

De Strzelecki, P. E. 1845. *A Physical Description of New South Wales and Van Diemen's Land.* London: Longmans.

Diamond, J. 1993. Ten thousand years of solitude. *Discover* 14(3): 48–57.

Diamond, J. 1997. *Guns, Germs, and Steel.* London: Chatto & Windus.

Diamond, J. 1998. The evolution of guns and germs. In A. C. Fabian (ed.), *Evolution: Society, Science and the Universe*: 46–63. Cambridge: Cambridge University Press.

Dibble, H. 1995. Middle Paleolithic scraper reduction: Background, clarification, and review of the evidence to date. *Journal of Archaeological Method and Theory* 2: 299–368.

Dobres, M., and Hoffman, C. 1994. Social agency and the dynamics of prehistoric technology. *Journal of Archaeological Method and Theory* 1: 211–258.

Donald, M. 1991. *The Origins of the Modern Mind: Three Stages in the Evolution of Culture and Cognition*. Cambridge, MA: Harvard University Press.

Dorion, R. (ed.). 2004. *Bitemark Evidence: A Color Atlas and Text*. New York: Dekker.

Douglas, M. 1975. *Implicit Meanings*. London: Routledge & Kegan Paul.

Dunbar, R. 2007. Why are humans not just great apes? In C. Pasternak (ed.), *What Makes Us Human?*: 37–48. Oxford: Oneworld.

Duyker, E. (ed.). 1992. *The Discovery of Tasmania: Journal Extracts from the Expeditions of Abel Janszoon Tasman and Marc-Joseph Marion Dufresne 1642 & 1772*. Hobart: St. David's Park Publishing/Tasmanian Government Printing Office.

Edgar, H. J. H., and Hunley, K. L. 2009. Race reconciled? How biological anthropologists view human variation. *American Journal of Physical Anthropology* 139: 1–4.

Edmonds, M. 1992. Their use is wholly unknown. In N. Sharples and A. Sheridan (eds.), *Vessels for the Ancestors: Essays on the Neolithic of Britain and Ireland in Honour of Audrey Henshall*: 179–193. Edinburgh: Edinburgh University Press.

Ehrenberg, M. 1989. *Women in Prehistory*. London: British Museum Publications.

Eliot, G. 1867. *Letters*, vol. 4. New Haven: Yale University Press.

Fagan, B. 1997. *Clash of Cultures*. Lanham: Alta Mira.

Fagan, B. M. (ed.). 2004. *The Seventy Great Inventions of the Ancient World*. London: Thames & Hudson.

Falk, D. 1997. Brain evolution in females. In L. D. Hager (ed.), *Women in Human Evolution*: 114–136. London: Routledge.

Falk, D. 2005. Prelinguistic evolution in early hominins: Whence motherese? *Behavioural and Brain Sciences* 27: 491–541.

Falk, D. 2009. *Finding Our Tongues: Mothers, Infants & the Origins of Languages*. New York: Basic Books.

Fedigan, L. M. 1997. Is primatology a feminist science? In L. D. Hager (ed.), *Women in Human Evolution*: 56–75. London: Routledge.

Flusser, V. 1999. *The Shape of Things: A Philosophy of Design*. London: Reaktion Books.

Foley, R., and Lahr, M. 1997. Mode 3 technologies and the evolution of modern humans. *Cambridge Archaeological Journal* 7(1): 3–36.

Frisch, M. *Homo Faber: Ein Bericht*. Frankfurt: Suhrkamp Verlag.

Gammon, K. S. 2007. Porn star auctions breast implants. *ABC News Medical Unit*, December 14.

Gell, A. 1992. The enchantment of technology and the technology of enchantment. In J. Coote and A. Shelton (eds.), *Anthropology, Art and Aesthetics*: 40–63. Oxford: Oxford University Press.

Gell, A. 1998. *Art and Agency: An Anthropological Theory*. Oxford: Clarendon Press.

Gellner, E. 1989. Culture, constraint and community: Semantic and coercive compensations for the genetic under-determination of Homo sapiens sapiens. In P. Mellars and C. Stringer (eds.), *The Human Revolution*: 17–30. Edinburgh: Edinburgh University Press.

Gibson, J. J. 1979. *The Ecological Approach to Visual Perception*. Boston: Houghton Mifflin.

Gibson, K. R., and Ingold, T. (eds.). 1993. *Tools, Language and Cognition in Human Evolution*. Cambridge and New York: Cambridge University Press.

Godelier, M. 1984. *L'idéel et le matériel*. Paris: Fayard.

Gomes, R., Levison, H. F., Tsiganis, K., and Morbidelli, A. 2005. Origin of the cataclysmic Late Heavy Bombardment period of the terrestrial planets. *Nature* 435: 466.

Goody, J. 1979. *The Domestication of the Savage Mind*. Cambridge: Cambridge University Press.

Gosden, C. 1994. *Social Being and Time*. Oxford: Blackwell.

Gosden, C. 2004. Towards a theory of material engagement. In E. DeMarrais, C. Gosden, and C. Renfrew (eds.), *Rethinking Materiality: The Engagement of Mind with the Material World*: 33–40. Cambridge: McDonald Institute for Archaeological Research.

Götherström, A., et al. 2005. Cattle domestication in the Near East was followed by hybridization with aurochs bulls in Europe. *Proceedings of the Royal Society B* 272(1579): 2345–2350.

Gott, B. 2002. Fire-making in Tasmania: Absence of evidence is not evidence of absence. *Current Anthropology* 43: 650–656.

Gould, S. J. 1977. *Ontogeny and Phylogeny*. Cambridge, MA: Belknap Press.

Gould, S. J. 1979. Mickey Mouse meets Konrad Lorenz. *Natural History* 88 (May): 30–36.

Graham, D. W. 1999. Empedocles and Anaxagoras: Responses to Parmenides. In A. A. Long (ed.), *The Cambridge Companion to Early Greek Philosophy*: 159–180. Cambridge: Cambridge University Press.

Graves-Brown, P. (ed.). 2000. *Matter, Materiality and Modern Culture*. London: Routledge.

Gray, J. 2002. *Straw Dogs: Thoughts on Humans and Other Animals*. London: Granta.

Guy, F., Lieberman, D. E., Pilbeam, D., Ponce de Leon, M. S., Likius, A., Mackaye, H. T., Vignaud, P., Zollikofer, C. P. E., and Brunet, M. 2006. Morphological affinities of the *Sahelanthropus tchadensis* (Late Miocene hominid from Chad) cranium. *Proceedings of the National Academy of Sciences of the USA (PNAS)* 102(52): 18836–18841.

Haesaerts, P., Damblon, F., Bachner, M., and Trnka, G. (1996). Revised stratigraphy and chronology of the Willendorf II sequence, Lower Austria. *Archaeologia Austriaca* 80: 25–42.

Hager, L. D. (ed.). 1997. *Women in Human Evolution*. London and New York: Routledge.

Hart, D., and Sussman, R.W. 2008. *Man the Hunted: Primates, Predators, and Human Evolution*. Boulder: Westview Press.

Heinrich, J. 2004. Demography and cultural evolution: How adaptive cultural processes can produce maladaptive losses—the Tasmanian case. *American Antiquity* 69: 197–214.

Henshilwood, C. S. 2008. Winds of change: Paleoenvironments, material culture and human behaviours in the late Pleistocene (~77 ka–48 ka ago) in the Western Cape province, South Africa. *South African Archaeological Society Goodwin Series* 10: 35–51.

Hill, K., Goodall, J., Pusey, A., Williams, J., Boesch, C., Boesch, H., and Wrangham, R. 2001. Mortality rates among wild chimpanzees. *Journal of Human Evolution* 40: 437–450.

Hiscock, P. 2006. Blunt and to the point: Changing technological strategies in Holocene Australia. In I. Lilley (ed.), *Archaeology of Oceana: Australia and the Pacific Islands*: 69–95. Oxford: Blackwell.

Hiscock, P. 2008. *The Archaeology of Ancient Australia*. London: Routledge.

Hiscock, P., and Attenbrow, V. 1998. Early Holocene backed artefacts from Australia. *Archaeology in Oceana* 33: 49–63.

Hladik, C. M., Chivers, D. J., and Pasquet, P. 1999. On diet and gut size in non-human primates and humans: Is there a relationship to brain size? *Current Anthropology* 40: 695–697.

Hodder, I. (ed.). 1987. *The Archaeology of Contextual Meanings*. Cambridge: Cambridge University Press.

Hodge, J., and Radick, G. (eds.). 2009. *The Cambridge Companion to Darwin* (2nd ed.). Cambridge and New York: Cambridge University Press.

Houellebecq, M. 2005. *La possibilité d'une île*. Paris: Fayard.

Hoyle, F., and Wickramasinghe, C. 1981. *Evolution from Space: A Theory of Cosmic Creationism*. London: Dent.

Hughes, R. 1987. *The Fatal Shore*. London: Collins Harvill.

Hurd, P. L., and van Anders, S. M. 2007. Latitude, digit ratios, and Allen's and Bergmann's rules: A comment on Loehlin, McFadden, Medland, and Martin (2006). *Archives of Sexual Behavior* 36: 139–141.

Ingold, T. 1998. The evolution of society. In A. C. Fabian (ed.), *Evolution: Society, Science and the Universe*: 79–99. Cambridge: Cambridge University Press.

Ingold, T. 2000. Making culture and weaving the world. In P. Graves-Brown (ed.), *Matter, Materiality and Modern Culture*: 50–71. London: Routledge.

Ingold, T. 2001a. From complementarity to obviation: On dissolving the boundaries between social and biological anthropology, archaeology and psychology. In S. Oyama, P. E. Griffiths, and R. D. Gray (eds.), *Cycles of Contingency: Developmental Systems and Evolution*: 225–279. Cambridge, MA: MIT Press.

Ingold, T. 2001b. Beyond art and technology: The anthropology of skill. In M. B. Schiffer (ed.), *Anthropological Perspectives on Technology* (Amerind Foundation New World Studies 5): 17–31. Albuquerque: University of New Mexico Press.

Ingold, T. 2007. Materials against materiality. *Archaeological Dialogues* 14: 1–16.

Jaynes, J. 1976. *The Origin of Consciousness in the Breakdown of the Bicameral Mind*. Boston: Houghton Mifflin.

Johanson, D. C., and Maitland, E. 1981. *Lucy: The Beginning of Humankind*. St. Albans: Granada.

Johanson, D. C. 2009. Lucy (*Australopithecus afarensis*). In M. Ruse and J. Travis (eds.), *Evolution: The First Four Billion Years*: 693–697. Cambridge, MA: Belknap Press.

Johnson, M. 2002. *Behind the Castle Gate: From Medieval to Renaissance*. London: Routledge.

Jones, A. 2004. Archaeometry and materiality: Materials-based analysis in theory and practice. *Archaeometry* 46(3): 327–338.

Jones, A. 2007. *Memory and Material Culture*. Cambridge: Cambridge University Press.

Jones, A., and MacGregor, G. 2002. *Colouring the Past: The Significance of Colour in Archaeological Research*. Oxford: Berg.

Jones, R. 1972. Tasmanian Aborigines and dogs. *Mankind* 7: 256–271.

Jones, R. 1977a. The Tasmanian paradox. In R. V. S. Wright (ed.), *Stone Tools as Cultural Markers: Change, Evolution and Complexity*: 189–204. Canberra: Australian Institute of Aboriginal Studies.

Jones, R. 1977b. Man as an element of a continental fauna: The case of the sundering of the Bassian bridge. In J. Allen, J. Golson, and R. Jones (eds.), *Sunda and Saul: Prehistoric Studies in Southeast Asia, Melanesia and Australia*: 317–386. London: Academic Press.

Jones, R. 1978. Why did the Tasmanians stop eating fish? In R. Gould (ed.), *Explorations in Ethnoarchaeology*: 11–48. Albuquerque: University of New Mexico Press.

Jones, R. 1995. Tasmanian archaeology: Establishing the sequences. *Annual Review of Anthropology* 24: 423–446.

Karlinsky, S. 1973. *Letters of Anton Chekhov*. London: Bodley Head.

Kauffman, S. 2004. Prolegomenon to a general biology. In W. A. Dembski and M. Ruse (eds.), *Debating Design: From Darwin to DNA*: 151–172. Cambridge: Cambridge University Press.

Kelly, K. 1994. *Out of Control: The New Biology of Machines*. London: Fourth Estate.

Kelly, K. 2007. The technium and the 7th kingdom of life. *Edge*, July 18. www.edge.org/documents/archive/edge217.html.

Kennedy, M. 2003. Kill looters, urges archaeologist. *The Guardian*, July 9.

Kern, A., Kowarik, K., Rausch, A. W., and Reschreiter, H. (eds.). 2008. *Salz-Reich: 7000 Jahre Hallstatt*. Vienna: NHM.

Kitcher, P. 2007. *Living with Darwin: Evolution, Design, and the Future of Faith*. Oxford and New York: Oxford University Press.

Kivell, T. L., and Schmitt, D. 2009. Independent evolution of knuckle-walking in African apes shows that humans did not evolve from a knuckle-walking ancestor. *Proceedings of the National Academy of Sciences of the USA* (*PNAS*) 106(34): 14241–14246.

Klein, R. G., with Edgar, B. 2002. *The Dawn of Human Culture*. New York: Wiley.

Knappett, C. 2002. Photographs, skeuomorphs and marionettes: Some thoughts on mind, agency and object. *Journal of Material Culture* 7: 97–117.

Knappett, C. 2004. The affordances of things: A post-Gibsonian perspective on the relationality of mind and matter. In E. DeMarrais, C. Gosden, and C. Renfrew (eds.), *Rethinking Materiality: The Engagement of Mind with the Material World*: 43–51. Cambridge: Cambridge University Press.

Knappett, C. 2005. *Thinking Through Material Culture: An Interdisciplinary Approach*. Pennysylvania: University of Pennsylvania Press.

Knappett, C. 2008. The neglected networks of material agency: Artefacts, pictures and texts. In C. Knappett and L. Malafouris (eds.), *Material Agency: Towards a Non-Anthropocentric Approach*: 139–156. New York: Springer.

Knappett, C., and Malafouris, L. (eds.). 2008. *Material Agency: Towards a Non-Anthropocentric Approach*. New York: Springer.

Knüsel, C. J. 1992. The throwing hypothesis and hominid origins. *Human Evolution* 7(1): 1–7.

Köhler, M., and Moyà-Solà, S. 1997. Ape-like or hominid-like? The positional behaviour of *Oreopithecus bambolii* reconsidered. *Proceedings of the National Academy of Sciences of the USA* (*PNAS*) 94: 11747–11750.

Kopytoff, I. 1986. The cultural biography of things: Commoditization as process. In A. Appadurai (ed.), *The Social Life of Things*: 64–94. Cambridge: Cambridge University Press.

Kurzweil, R. 2005. *The Singularity Is Near: When Humans Transcend Biology*. London: Duckworth.

Laland, K. N., and Odling-Smee, J. 2000. The evolution of the meme. In R. Augner (ed.), *Darwinizing Culture: The Status of Memetics as a Science*: 121–142. Oxford: Oxford University Press.

Laland, K. N., Odling-Smee, J., and Myles, S. 2010. How culture shaped the human genome: Bringing genetics and the human sciences together. *Nature Reviews: Genetics* 11: 137–148.

Latour, B. 1999. *Pandora's Hope: Essays on the Reality of Science Studies*. Cambridge, MA: Harvard University Press.

Leakey, M. D., and Hay, R. L. 1979. Pliocene footprints in the Laetoli Beds at Laetoli, northern Tanzania. *Nature* 278: 317–323.

Leeuw, S. E. van der. 1994. Cognitive aspects of "technique." In C. Renfrew and E. B. W. Zubrow (eds.), *The Ancient Mind, Elements of Cognitive Archaeology*: 135–142. Cambridge: Cambridge University Press.

Lemonnier, P. 1986. The study of material culture today: Toward an anthropology of technical systems. *Journal of Anthropological Archaeology* 5: 147–186.

Lemonnier, P. 1992. *Elements for an Anthropology of Technology* (University of Michigan Museum of Anthropology, Anthropological Paper No. 88). Michigan: University of Michigan Museum of Anthropology.

Leroi-Gourhan, A. 1964 [1943–1946]. *Le Geste et la parole*, vol. 1: *Technique et langage*; vol. 2: *La Mémoire et les rythmes*. Paris: Albin Michel.

Leroux, B. 2007. *Recettes de Normandie*. Caen: Éditions Ouest-France.

Lévi-Strauss, C. 1970. *The Raw and the Cooked*. New York: Harper & Row.

Lewin, R. 1997. *Bones of Contention: Controversies in the Search for Human Origins* (2nd ed.). Chicago: University of Chicago Press.

Lewin, R., and Foley, R. A. 2004. *Principles of Human Evolution*. Oxford: Blackwell.

Lieberman, D. E., McBratney, B. M., and Krovitz, G. 2002. The evolution and development of cranial form in *Homo sapiens*. *Proceedings of the National Academy of Sciences of the USA (PNAS)* 99(3): 1134–1139.

Lieberman, L. 2001. How "Caucasoids" got such big crania and why they shrank. *Current Anthropology* 42(1): 69–95.

Lillios, K. 2003. Creating memory in prehistory: The engraved slate plaques of southwest Iberia. In R. M. Van Dyke and S. E. Alcock (eds.), *Archaeologies of Memory*: 129–150. Oxford: Blackwell.

Lordkipanidze, D., Vekua, A., Ferring, R., Rightmire, G. P., Zollikofer, C.P.E., Ponce De León, M. S., Agusti, J., Kiladze, G., et al. 2006. A fourth hominin skull from Dmanisi, Georgia. *The Anatomical Record Part A: Discoveries in Molecular, Cellular, and Evolutionary Biology* 288(11): 1146–1157.

Lourandos, H. 1983. 10,000 years in the Tasmanian highlands. *Australian Archaeology* 16: 39–47.

Lourandos, H. 1985. Intensification and Australian prehistory. In T. D. Price and J. A. Brown (eds.), *Prehistoric Hunter-Gatherers: The Emergence of Cultural Complexity*: 385–423. Orlando, FL: Academic Press.

Lourandos, H. 1997. *Continent of Hunter-Gatherers: New Perspectives in Australian Prehistory*. Cambridge: Cambridge University Press.

Lovell, N. C., Jurmain, R., and Kilgore, L. 2000. Skeletal evidence of probable treponemal infection in free-ranging African apes. *Primates* 41(3): 275–290.

Lynch, G., and Granger, R. 2008. *Big Brain: The Origins and Future of Human Intelligence*. New York: Palgrave Macmillan.

Malafouris, L. 2004. The cognitive basis of material engagement: Where brain, body and culture conflate. In E. DeMarrais, C. Gosden, and C. Renfrew (eds.), *Rethinking Materiality: The Engagement of Mind with the Material World*: 53–62. Cambridge: Cambridge University Press.

Maslin, M. A., and Christensen, B. 2007. Tectonics, orbital forcing, global climate change, and human evolution in Africa: Introduction to the African paleoclimate. *Journal of Human Evolution* 53: 443–464. Special volume.

Mason, I. L. 1996. *A World Dictionary of Livestock Breeds, Types and Varieties* (4th ed.). C.A.B. International.

Mauss, M. 1990. *The Gift: The Form and Reason for Exchange in Archaic Societies* (translated by W. D. Halls from *Essai sur le Don*, 1950) London: Routledge.

Mauss, M. 2004. *Marcel Mauss: Techniques, Technology and Civilisation* (edited and introduced by N. Schlanger). Oxford: Bergham Press.

McGrew, W. C. 1987. Tools to get food: The subsistants of Tasmanian Aborigines and Tanzanian chimpanzees compared. *Journal of Anthropological Research* 43: 247–258.

McKiernan, C., and Lieberman, S. A. 2005. Circulatory shock in children. *Pediatrics in Review* 26: 451–460.

McLuhan, M. 1964. *Understanding Media: The Extensions of Man*. London: Routledge.

Mellars, P. A., and Gibson, K. (eds.). 1996. *Modelling the Early Human Mind*. Cambridge: McDonald Institute for Archaeological Research.

Meltzer, D., and Holliday, V. T. Would North American Paleoindians have noticed Younger Dryas Age climate changes? *Journal of World Prehistory* 23(1).

Meskell, L. 2004. Divine things. In E. DeMarrais, C. Gosden, and C. Renfrew (eds.), *Rethinking Materiality: The Engagement of Mind with the Material World*: 249–259. Cambridge: Cambridge University Press.

Mesoudi, A., Whiten, A., and Laland, K. N. 2004. Perspective: Is human cultural evolution Darwinian? Evidence reviewed from the perspective of *The Origin of Species*. *Evolution* 58(1): 1–11.

Midgley, M. 2003. *The Myths We Live By*. London: Routledge.

Miller, D. 1985. *Artefacts as Categories: A Study of Ceramic Variability in Central India*. Cambridge: Cambridge University Press.

Miller, D. 1987. *Material Culture and Mass Consumption*. Oxford: Blackwell.

Miller, D. 2002. Coca-Cola: A black sweet drink from Trinidad. In V. Buchli (ed.), *The Material Culture Reader*: 245–264. New York: Berg.

Miller, D. (ed.). 2005. *Materiality*. Durham, NC: Duke University Press.

Mithen, S. 1994. From domain specific to generalized intelligence: A cognitive interpretation of the Middle/Upper Paleolithic transition. In C. Renfrew and E. B. W. Zubrow (eds.), *The Ancient Mind, Elements of Cognitive Archaeology*: 29–39. Cambridge: Cambridge University Press.

Mithen, S. 1996. *The Prehistory of the Mind: A Search for the Origin of Art, Religion and Science*. London: Orion.

Mithen, S. (ed.). 1998. *Creativity in Human Evolution and Prehistory*. London and New York: Routledge.

Mollison, T. 1994. The eloquent bones of Abu Hureyra. *Scientific American* 271(2): 60–65.

Morris, D. 1967. *The Naked Ape: A Zoologist's Study of the Human Animal*. London: Jonathan Cape.

Morton, J. B. 1994. *Cram Me with Eels!* London: Methuen.

Mullen, W. 2005. Gorilla attacks keeper at zoo. *Chicago Tribune*, July 6. www.chicagotribune.com/chi-0507060240jul06,0,4171751.story.

Müller, W., Fricke, H., Halliday, A. N., McCulloch, M. T., and Wartho, J.-A. 2003. Origin and migration of the alpine iceman. *Science* 302 (5646): 862–866.

Mulvaney, J. 2007. *The Axe Had Never Sounded: Place, People and Heritage in Recherche Bay, Tasmania* (Aboriginal History Monograph, 14). Canberra; Australian National University.

Oppenheimer, S. 2007. What makes us human?—Our ancestors and the weather. In C. Pasternak (ed.), *What Makes Us Human?*: 93–113. Oxford: Oneworld.

Paley, W. 1837. *Natural Theology: Or, Evidences of the Existence and Attribute of the Deity, Collected from the Appearances of Nature*. Boston: Gould, Kendall and Lincoln.

Paley, W. 1996. *Natural Theology* (introduction and notes by M. D. Eddy and D. M. Knight). Oxford University Press.

Pasternak, C. (ed.). 2007. *What Makes Us Human?* Oxford: Oneworld.

Paterson, H. 1985. The recognition concept of species. In E. Vrba (ed.), *Species and Speciation* (Transvaal Museum Monograph 4): 21–29. Pretoria: Transvaal Museum.

Paul, D. B. 2009. Darwin, social Darwinism and eugenics. In J. Hodge and G. Radick (eds.), *The Cambridge Companion to Darwin* (2nd ed.): 219–245. Cambridge and New York: Cambridge University Press.

Pfaffenberger, B. 1988. Fetishised objects and humanised nature: Toward an anthropology of technology. *Man* 23: 236–252.

Pfaffenberger, B. 1992. Technological dramas. *Science, Technology, & Human Values* 17(3): 282–312.

Pfeiffer, J. E. 1982. *The Creative Explosion*. New York: Harper & Row.

Piontelli, A. 2008. *Twins in the World: The Legends They Inspire and the Lives They Lead*. New York: Palgrave Macmillan.

Portmann, A. 1941. Die Tragzeiten der Primaten und die Dauer der Schwangerschaft beim Menschen: Ein Problem der vergleichenden Biologie. *Revue suisse de Zoologie* 48: 511–518.

Postgate, J. N. 1994. Text and figure in ancient Mesopotamia: Match and mismatch. In C. Renfrew and E. B. W. Zubrow (eds.), *The Ancient Mind: Elements of Cognitive Archaeology*: 176–184. Cambridge: Cambridge University Press.

Prat, S. 2007. The quaternary boundary: 1.8 or 2.6 million years old? Contributions of early *Homo*. *Revue de l'Association française pour l'étude du Quaternaire* 18(1): 99–107.

Pratchett, T. 1988. *Wyrd Sisters*. London: Corgi.

Premo, L. S., and Hublin, J.-J. 2009. Culture, population structure, and low genetic diversity in Pleistocene hominins. *Proceedings of the National Academy of Sciences of the USA (PNAS)* 106(1): 33–37.

Purves, D., and Lichtman, J. W. 1985. *Principles of Neural Devolopment*. Sunderland, MA: Sinauer Associates.

Read, D. 2006. Tasmanian knowledge and skill: Maladaptive imitation or adaptive technology. *American Antiquity* 71: 164–184.

Renfrew, C. 1982. *Towards an Archaeology of Mind (Inaugural Lecture)*. Cambridge: Cambridge University Press.

Renfrew, C. 1994. Towards a cognitive archaeology. In C. Renfrew and E. B. W. Zubrow (eds.), *The Ancient Mind, Elements of Cognitive Archaeology*: 3–12. Cambridge: Cambridge University Press.

Renfrew, C. 2001. Symbol before concept. In I. Hodder (ed.), *Archaeological Theory Today*: 122–140. London: Polity Press.

Renfrew, C. 2003. *Figuring It Out*. London: Thames & Hudson.

Renfrew, C. 2004. Towards a theory of material engagement. In E. DeMarrais, C. Gosden, and C. Renfrew (eds.), *Rethinking Materiality: The Engagement of Mind with the Material World*: 23–31. Cambridge: McDonald Institute Monographs.

Renfrew, C., and Scarre, C. (eds.). 1998. *Cognition and Material Culture: The Archaeology of Symbolic Storage*. Cambridge: McDonald Institute for Archaeological Research.

Renfrew, C., and Zubrow, E. B. W. (eds.). 1994. *The Ancient Mind, Elements of Cognitive Archaeology*. Cambridge: Cambridge University Press.

Richards, M. P., and Hedges, R. E. M. 2000. Focus: Gough's Cave and Sun Hole Cave human stable isotope values indicate a high animal protein diet in the British Upper Paleolithic. *Journal of Archaeological Science* 27: 1–3.

Richards, M. P., Molleson, T. I., Martin, L., Russell, N., and Pearson, J. A. 2003. Palaeodietary evidence from Neolithic Çatalhöyük, Turkey. *Journal of Archaeological Science* 30: 67–76.

Richerson, P. J., Boyd, R., and Bettinger, R. L. 2001. Was agriculture impossible during the Pleistocene but mandatory during the Holocene? *American Antiquity* 66: 387–412.

Roaf, M. 1990. *Cultural Atlas of Mesopotamia and the Ancient Near East*. New York: Facts on File.

Robb, J. 2004. The extended artefact and the monumental economy: A methodology for material agency. In E. DeMarrais, C. Gosden, and C. Renfrew (eds.), *Rethinking Materiality: The Engagement of Mind with the Material World*: 131–139. Cambridge: McDonald Institute for Archaeological Research.

Rollo, F., Ubaldi, M., Ermini, L., and Marota, I. 2002. Ötzi's last meals: DNA analysis of the intestinal content of the Neolithic glacier mummy from the Alps. *Proceedings of the National Academy of Sciences of the USA (PNAS)* 99: 12594–12599.

Rozoy, J.-G. 2003. The evolution of the brain is still progressing. *L'Anthropologie* 107(5): 645–687.

Ruff, C. B. (1994). Morphological adaptation to climate in modern and fossil hominids. *Yearbook of Physical Anthropology* 37: 65–107.

Ruff, C. 1995. Biomechanics of the hip and birth in early *Homo*. *American Journal of Physical Anthropology* 98: 527–574.

Ruff, C. 2002. Variation in human body size and shape. *Annual Review of Anthropology* 31: 211–232.

Ruff, C. 2010. Body size and body shape in early hominins—implications of the Gona pelvis. *Journal of Human Evolution* 58(2): 166–170.

Ruff, C. B., Holt, B. M., Sládek, V., Berner, M., Murphy, W. A., Nedden, D. zur, Seidler, H., and Recheis, W. 2006. Body size, body proportions, and mobility in the Tyrolean "Iceman." *Journal of Human Evolution* 51(1): 91–101.

Sanger, M. 1920. *What Every Girl Should Know*. Springfield, IL: United Sales Co.

Sartre, J.-P. 1976. *Critique of Dialectical Reason* (translated by A. Sheridan-Smith). London: New Left Books.

Scarre, C. (ed.). 2009. *The Human Past: World Prehistory and the Development of Human Societies* (2nd ed.). London: Thames & Hudson.

Schaefer, E. 2002. Gauging the revolution: 16mm film and the rise of the pornographic feature. *Cinema Journal* 41(3): 3–26.

Schidlowski, M. 2001. Carbon isotopes as biogeochemical recorders of life over 3.8 Ga of Earth history: Evolution of a concept. *Precambrian Research* 106: 117–134.

Schiffer, M. B. 2001. Toward an anthropology of technology. In M. B. Schiffer (ed.), *Anthropological Perspectives on Technology* (Amerind Foundation New World Studies 5): 1–15. Albuquerque: University of New Mexico Press.

Schlanger, N. 1994. Mindful technology: Unleashing the *chaîne opératoire* for an archaeology of mind. In C. Renfrew and E. B. W. Zubrow (eds.), *The Ancient Mind, Elements of Cognitive Archaeology*: 143–151. Cambridge: Cambridge University Press.

Schlanger, N. 1996. Understanding Levallois: Lithic technology and cognitive archaeology. *Cambridge Archaeological Journal* 6(2): 231–254.

Secord, J. 2008. Introduction. In C. Darwin, *Evolutionary Writings* (edited by J. A. Secord): vii–xlviii. Oxford and New York: Oxford University Press.

Semaw, S., Rogers, M. J., Quade, J., Renne, P. R., Butler, R. F., Dominguez-Rodrigo, M., Stout, D., Hart, W. S.,Travis Pickering, T., and Simpson, S. W. 2003. 2.6-million-year-old stone tools and associated bones from OGS-6 and OGS-7, Gona, Afar, Ethiopia. *Journal of Human Evolution* 45(2): 169–177.

Semenov, P. P. 1998. *Travels in the Tian'-Shan' 1856-1857* (edited by C. Thomas). London: Hakluyt Society.

Senut, B., Pickford, M., Gommery, D., Mein, P., Cheboi, K., and Coppens, Y. 2001. First hominid from the Miocene (Lukeino Formation, Kenya). *Comptes Rendus de l'Académie des Sciences* (Series 11A): *Earth and Planetary Science* 332(2): 137–144.

Shennan, S. J. 2002. *Genes, Memes and Human History: Darwinian Archaeology and Cultural Evolution*. London: Thames & Hudson.

Sherratt, A., and Taylor, T. 1997. Metal vessels in Bronze Age Europe and the context of Vulchetrun. In A. Sherratt (ed.), *Economy and Society in Prehistoric Europe*: 431–456. Edinburgh: Edinburgh University Press.

Short, P. 2003. *In Pursuit of Plants: Experiences of Nineteenth & Early Twentieth Century Plant Collectors*. Portland: Timber Press.

Sim, R. 1999. Why the Tasmanians stopped eating fish: Evidence for the late Holocene expansion in resource exploitation strategies. In J. Hall and I. J. McNiven (eds.), *Australian Coastal Archaeology*: 263–269. Canberra: Australian National University.

Soffer, O., Adovasio, J. M., and Hyland, D. C. 2000. The "Venus" figurines: Textiles, basketry, gender, and status in the Upper Paleolithic. *Current Anthropology* 41(4): 511–538.

Soffer, O., Adovasio, J. M., and Hyland, D. C. 2001. Perishable technologics and invisible people: Nets, baskets, and "Venus" wear ca. 26,000 BP. In B. Purdy (ed.), *Enduring Records: The Environmental and Cultural Heritage of Wetlands*: 233–245. Oxford: Oxbow Books.

Spencer, H. 1864. *Principles of Biology*. London: Williams and Norgate.

Sperber, D. 1992. Culture and matter. In J.-C. Gardin and C. S. Peebles (eds.), *Representations in Archaeology*: 56–65. Bloomington and Indianapolis: Indiana University Press.

Sperber, D. 1994. The modularity of thought and the epidemiology of representations. In L. A. Hirschfeld and S. A. Gelman (eds.), *Mapping the Mind*: 39–67. Cambridge: Cambridge University Press.

Sperber, D. 2000. An objection to the memetic approach to culture. In R. Aunger (ed.), *Darwinizing Culture: The Status of Memetics as a Science*: 163–174. Oxford: Oxford University Press.

Spindler, K. 1994. *The Man in the Ice*. London: Weidenfeld & Nicolson.

Sponheimer, M., and Lee-Thorp, J. 2003. Differential resource utilization by extant great apes and australopithecines: Towards solving the C_4 conundrum. *Comparative Biochemistry and Physiology—Part A: Molecular & Integrative Physiology* 136(1): 27–34.

Stark, F. 1997. *Baghdad Sketches*. Marlboro, VT: Marlboro Press.

Steudel-Numbers, K. L., Weaver, T. D., and Wall-Scheffler, C. M. 2007. The evolution of human running: Effects of changes in lower-limb length on locomotor economy. *Journal of Human Evolution* 53: 191–196.

Stringer, C., and McKie, R. 1997. *African Exodus: The Origins of Modern Humanity*. London: Pimlico.

Sutherland, A. 1888. *Victoria and Its Metropolis: Past & Present* (2 vols.). Melbourne: McCarron, Bird & Co.

Sutton, J. 2008. Material agency, skills and history: Distributed cognition and the archaeology of memory. In C. Knappett and L. Malafouris (eds.), *Material Agency: Towards a Non-Anthropocentric Approach*: 37–55. New York: Springer.

Tacon, P. 1994. Socialising landscapes: The long-term implications of signs, symbols and marks on the land. *Archaeology in Oceana* 29: 117–129.

Tattersall, I. 1998. *Becoming Human: Evolution and Human Uniqueness*. Oxford: Oxford University Press.

Tattersall, I. 2007. Human evolution and the human condition. In C. Pasternak (ed.), *What Makes Us Human?*: 133–145. Oxford: Oneworld.

Taylor, E. B. 1893. On the Tasmanians as representative of Paleolithic man. *Journal of the Anthropological Institute* 23: 141–152.

Taylor, T. 1996. *The Prehistory of Sex: Four Million Years of Human Sexual Culture*. London: 4th Estate.

Taylor, T. 1999. Envaluing metal: Theorizing the Eneolithic "hiatus." In S. Young, A. M. Pollard, P. Budd, and R. A. Ixer (eds.), *Metals in Antiquity*: 22–32. Oxford: Archaeopress.

Taylor, T. 2002. *The Buried Soul: How Humans Invented Death*. London: 4th Estate.

Taylor, T. 2006a. The human brain is a cultural artifact. In J. Brockman (ed.), *Edge 175: The EDGE Annual Question 2006* (www.edge.org).

Taylor, T. 2006b. Why the Venus of Willendorf has no face. *Archäologie Österreichs* 17(1): 26–29.

Taylor, T. 2006c. The evolution of human sexual culture. In M. Kauth (ed.), *Journal of Psychology & Human Sexuality* (special double issue on the "Evolution of Sexual Attraction"): chapter 3. Binghamton, NY: Haworth Press.

Taylor, T. 2008. The Willendorf Venuses: Notation, iconology and materiality. *Mitteilungen der Anthropologischen Gesellschaft in Wien (MAGW)* 138: 37–49.

Taylor, T. 2009. Inventing death. In E. Azoulay, A. Demian, and D. Frioux (eds.), *100,000 Years of Beauty: Prehistory/Foundations*: 128–131. Paris: Gallimard.

Trauth, M. H., Maslin, M. A., Deino, A. L., Strecker, M. R., Bergner, A. G. N., and Dühnforth, M. 2007. High- and low-latitude forcing of Plio-Pleistocene East African climate and human evolution. *Journal of Human Evolution* 53: 475–486.

Trigger, B. 2006. *A History of Archaeological Thought* (2nd ed.). Cambridge: Cambridge University Press.

Valoch, K. 2009. The Brno puppet. In E. Azoulay, A. Demian, and D. Frioux (eds.), *100,000 Years of Beauty: Prehistory/Foundations*: 180–183. Paris: Gallimard.

Venditti, C., Meade, A., and Pagel, M. 2010. Phylogenies reveal new interpretation of speciation and the Red Queen. *Nature* 463: 349–352.

Vickers, M., and Gill, D. 1994. *Artful Crafts: Ancient Greek Silverware and Pottery*. Oxford: Clarendon Press.

Viegas, J. 2000. "Venus wear" reveals world's oldest fashion. World's oldest hats discovered? ABCNEWS.com, May 9.

Vrba, E. 1996. Climate, heterochrony, and human evolution. *Journal of Anthropological Research* 52: 1–28.

Vygotsky, L. S. 1978. *Mind in Society: The Development of Higher Psychological Processes*. Cambridge, MA: Harvard University Press.

Vygotsky, L. S. 1986. *Thought and Language*. Cambridge, MA: Harvard University Press.

Wade, N. 2007. *Before the Dawn: Recovering the Lost History of Our Ancestors*. London: Duckworth.

Wallace, A. R. 1856. On the habits of the Orang-Utan of Borneo. *Annals & Magazine of Natural History* 18 (Series 2): 26–32. Wall-Scheffler, C. M., Geiger, K., and Steudel-Numbers, K. L. 2007. Infant carrying: The role of increased locomotory costs in early tool development. *American Journal of Physical Anthropology* 133: 841–846.

Weaver, T. D., and Steudel-Numbers, K. 2005. Does climate or mobility explain the differences in body proportions between Neandertals and their Upper Paleolithic successors? *Evolutionary Anthropology* 14: 218–223.

Weiss, A. E., Kroeger, T., Haney, J. C., and Fascione, N. 2009. Social and ecological benefits of restored wolf populations. *Transactions of the 72nd North American Wildlife and Natural Resources Conference*: 297–319.

White, L. 1959. *The Evolution of Culture*. New York: McGraw Hill.

White, T. D. 1986. Cut marks on the Bodo cranium: A case of prehistoric defleshing. *American Journal of Physical Anthropology* 69: 503–509.

Whiten, A., Goodall, J., McGrew, W. C., Nishida, T., Reynolds, V., Sugiyama, Y., Tutin, C. E. G., Wrangham, R. W., and Boesch, C. 1999. Cultures in chimpanzees. *Nature* 399: 682–685.

Wickelgren, I. 1999. Nurture helps mold able minds. *Science* 283: 1832–1834.

Wilford, J. N. 2007. Lost in a million year gap, solid clues to human origins. *New York Times*, September 18.

Wills, C. 1993. *The Runaway Brain: The Evolution of Human Uniqueness*. London: Harper Collins.

Wolpoff, M. H. 2009. How Neandertals inform human variation. *American Journal of Physical Anthropology* 139: 91–102.

Wood, B. 2006. Paleoanthropology: A precious little bundle. *Nature* 443: 278–281.

Wrangham, R. 2007. The cooking enigma. In C. Pasternak (ed.), *What Makes Us Human?*: 182–203. Oxford: Oneworld.

Wrangham, R. 2009. *Catching Fire: How Cooking Made Us Human*. London: Profile Books.

Zihlman, A. 1997. The paleolithic glass ceiling. In L. D. Hager (ed.), *Women in Human Evolution*: 91–113. London: Routledge.

Zilhão, J., Angelucci, D. E., Badal-Garcia, E., d'Errico, F., et al. 2010. Symbolic use of marine shells and mineral pigments by Iberian Neanderthals. *Proceedings of the National Academy of Sciences of the USA* (*PNAS*) [early edition: www.pnas.org/cg/doi/10.1073].

INDEX